WHERE HAVE ALL THE FLOWERS GONE?

CHARLES FLOWER

WHERE HAVE ALL THE FLOWERS GONE?

RESTORING WILD FLOWERS TO THE GARDEN AND COUNTRYSIDE

PHOTOGRAPHY BY MIKE BAILEY & STEVE WILLIAMS

The earth is the Lord's, and everything in it
I Corinthians, chapter 10, verse 26

To all those who have fought to save the best of our wildlife habitats, without whose efforts the process of restoration would have been almost impossible

Design Director: Alexandra Papadakis
Designer: Leyton Brown
Design Assistant: Hayley Williams
Editor: Sheila de Vallée

Front cover design: Leyton Brown

First published in Great Britain in 2008
reprinted in 2010
by PAPADAKIS PUBLISHER

PAPADAKIS

An imprint of New Architecture Group Ltd.

Kimber Studio, Winterbourne
Berkshire, RG20 8AN, UK
www.papadakis.net

ISBN: 978 1 901092 82 0

A CIP catalogue of this book is available from the British Library.

Printed in China

half title: Bluebells in a beech wood – this stunning sight can be created in a lifetime

frontis: Meadow cranesbill is susceptible to. herbicides and produces few seeds but once established it is magnificent

title: If we provide nectar, customers will turn up (in this case a Marbled White butterfly) – so long as we don't leave it too late

CONTENTS

CHAPTER 1
WHERE HAVE ALL THE FLOWERS GONE?

OUR COUNTRYSIDE USED TO BE FULL OF WILD FLOWERS. THE HAY MEADOWS WERE PARTICULARLY RICH, BECAUSE THE NEEDS OF THE FARMER SO CLOSELY MATCHED THOSE OF WILDLIFE, AND WILD FLOWERS IN PARTICULAR.

But the hedges, lanes, fields and pastures were also carpeted with wild flowers, as were the footpaths and the margins of field ponds.

The wild flowers provided the nectar to keep myriads of insects, including abundant butterflies on the wing. The medicinal properties of our wild flowers were well known and every parish had, so to speak, a well-stocked medicine cupboard. But it was the farming system that was so important because it had evolved in such a way that farm management was the same as wild flower and wildlife management. Farming was then small scale and local. Milk, beef and arable crops supplied local markets and the mixed farm, which was the norm, consisted of hay meadows, pasture, ponds, hedges and spinneys, all of which provided the infrastructure necessary for the farm.

After the Second World War farming began to intensify with the horse being replaced by the Ferguson tractor – the "grey Fergie" – and the first combine harvesters arrived. Mechanisation was beginning to make an impact but wild flowers were still ubiquitous and the mixed farm with its rotations and low chemical usage was still king. In the 1960s chemicals began to get a grip on farm practices with better methods of applying chemical fertiliser and the introduction of pesticides. Then, in 1973, Britain joined the Common Market. Nothing could

above: A huge arable field leaves no room for wildlife

opposite: Everyone tried to pretend at the time that huge cornfields had little impact on wildlife, but declines of 50% or more in populations of common farmland birds and butterflies tell the real story

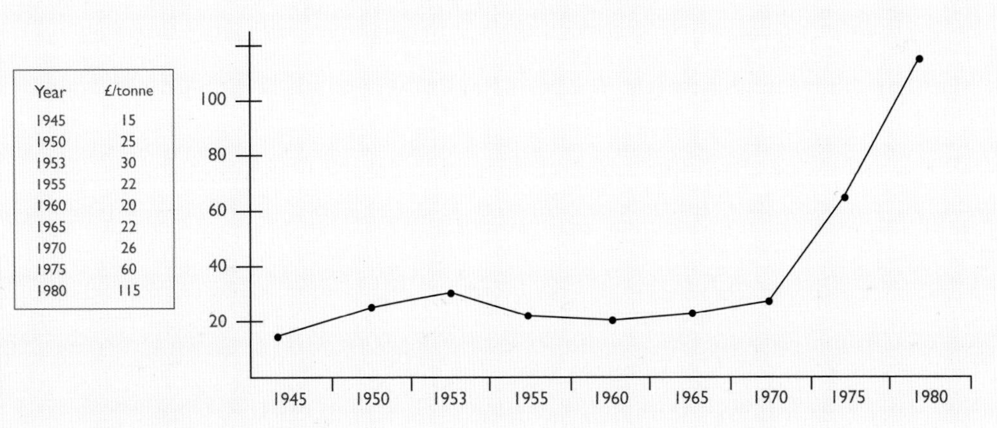

Year	£/tonne
1945	15
1950	25
1953	30
1955	22
1960	20
1965	22
1970	26
1975	60
1980	115

pages 6-7: We can only guess at the wild flowers that may have been abundant in the medieval fields. In this photograph the opportunist ox-eye daisy has spread into a red campion seed crop

have prepared us for what followed. Within nine years the price of corn, which had been £30 a ton for some time, doubled and redoubled to £120 a ton and suddenly even the most marginal land was being ploughed up to grow corn. The generous grants for increasing production, for hedge removal, drainage, and woodland clearance made matters worse. They were grants for the destruction of the countryside.

The County Wildlife Trusts, which are voluntary organisations to protect wildlife, had not been needed up to this point. Most had not even been formed. With this wholesale destruction there was a race against time to try and protect the best of our wildlife habitats, but the funds available were completely inadequate and only a tiny number were saved. In no time at all most of the hay meadows had been ploughed up and nearly all our wild flowers, with the butterflies and other insects that depended on them, had disappeared. So swift and so total was this destruction that today few people know anything about our heritage of wild flowers. In some counties where new landowners have recently been repairing the countryside structure of hedge, wood and pond and where they have asked for wild flowers to be restored, you will search in vain to find clues as to what might have been commonplace before the destruction. Along a nearby lane you may find some tufted vetch, a clump of knapweed, and maybe a couple of other species, a tiny fragment of the delights that were all around us a hundred years ago. Even if you have not read about wild flowers in the literature, a trip to parts of France, or Eastern Europe in particular, will raise questions in the mind. How could it possibly have happened?

In the case of momentous events such as these, there are always lessons to be learned, however painful. So I begin, reluctantly, with a brief description of this destruction. Of course it all started with the best of intentions – to grow more food from our own resources – but it turned out to be one of the better examples of the law of unintended consequences.

When a farm specialises as an arable unit, the animals go and the grass is ploughed up. Many of the smaller grass paddocks are merged to form larger fields to suit arable operations. The hedges and ponds of those smaller fields also disappear, as do any small woods or spinneys at the intersections of several hedges.

All the wild flowers associated with these habitats will be swept away, but there will be many other, less obvious casualties. Cornfields need to be levelled for the huge machines that cultivate at high speed, so it is not just the ponds that go, but all the features associated with them, the ditches and banks that provided differences in wetness and soil depth. The wild flowers that had colonised these different habitats over the years were replaced by just one species, corn.

But the saddest consequence of all was the ploughing up of the old hay meadows with their diversity of wild flowers, orchids and other wildlife. No one set out to do this, but with no animals to eat the hay, what was the farmer to do? Of course, with the benefit of hindsight the Ministry of Agriculture – as it was then called – might have advised on the retention of a representative collection of habitats in a given area, but this is wishful thinking. Massive subsidies cause massive changes. It was the usual sad story of nobody realising what was about to be lost until the losses had occurred. These meadows were the product of hundreds of years of intensive management dating from a time when feeding the densely populated Midland shire counties of medieval England was a major undertaking.

But there were other, more sinister effects of this intensification. With all the grass going under the plough, the long boom of the farm

Destruction of the Countryside

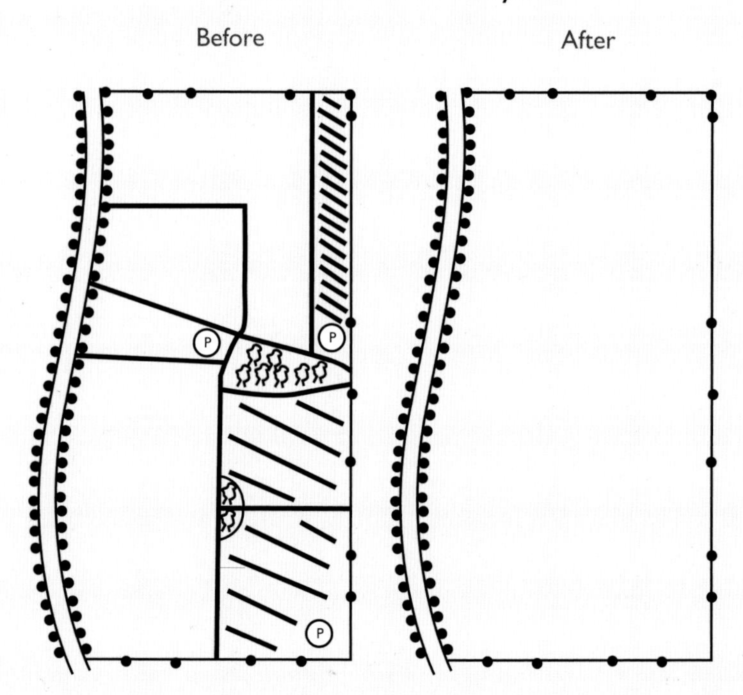

KEY	Change
Green lane with trees	nil
Hedge with occasional trees	- 57%
Woods	-100%
Pond	-100%
Hay meadow/permanent grass	-100%
Ley (temporary) grass	-100%
Arable	+ 65%

Farm simplification in order to increase corn production sounds reasonable until you count the cost in terms of habitat loss. When part of a mixed farm is turned into one 75-acre field, the outer boundary remains unchanged but the loss of hedges is more than 50 per cent and the woods, ponds and meadows are eliminated with all their associated species

Mixed farms with livestock, corn crops and varied habitat provide for a wide variety of wildlife

Flooded grassland shows up differences in ground levels where a wider range of soil conditions encourage a wider range of wild flowers

sprayer was attacking almost everywhere. Formerly there would have been many corners of the farm that were protected. The hay meadows would have seen a little farmyard manure and the grass pastures may have had a little nitrogen, but there would have been plenty of wild flowers, especially in the hedges. As soon as the farm was all arable, no corner was safe from an increasingly effective arsenal of chemicals.

One of the newest and most effective of these was a new total herbicide called "Round-Up," which killed all plants. Because corn prices were so high, farmers were tempted to eliminate all so-called weed opposition. Hedge bottoms were sprayed to stop "pernicious" weeds from invading the precious crop. What was in fact happening was that the stable perennial grass/wild flowers of the hedge were being damaged or destroyed by herbicides and, as a consequence, weeds such as barren brome, an annual that thrives on generous helpings of nitrogen fertiliser, were spreading into the fertile crop. This practice of spraying hedge bottoms, at one stage quite widespread, led to a situation encountered on many farms where there is hardly a single surviving wild flower apart from meadow buttercup, common sorrel and yarrow.

Once the all arable farm became the norm, the chemical onslaught on the countryside intensified for another reason. Until the end of the Second World War, corn farming had mostly been a springtime affair. Land was ploughed over winter, frosts helped to break it down to a workable seedbed, and sowing time was as early as the land would allow. With the advent of fungicides, the plant breeders soon

Herbicides get everywhere unless great care is taken

opposite: Great burnet in profusion at North Meadow Cricklade

The practice of spraying right up to hedges annihilated most wild flowers within a few years. A two-metre safety margin from the centre of the hedge has now been made mandatory

discovered that an autumn sown crop would give a bigger yield so long as the mildews and fungi could be kept at bay through the late autumn. Soon 90 per cent of arable crops were autumn sown and another deluge of chemicals descended on the countryside. Indeed, so often were these chemicals applied that tramlines were left unsown to guide the sprayer through the crop. This is the chemical signature on our farms today.

There were two other massive wildlife casualties of this practice: the winter stubbles – a source of so much winter food for finches, buntings, tree sparrows and many others – were removed; and the spring crops, which were the right height for lapwings and skylarks were replaced by autumn crops that were too tall in the spring for the nesting birds to see predators such as the fox. As a result of these changes together with the loss of habitat, populations of these once common farmland birds have declined by 50 per cent or more.

The effects of chemicals were also felt on grass farms. One subject only recently understood and still rarely discussed concerns the modern wormers given to horses and farm livestock. Many of these wormers are persistent in the dung and adversely affect the larvae of dung flies, beetles and other dung-feeding invertebrates. As a result species such as the hornet robber fly are now nationally scarce and countless other insects have declined dramatically. Yet where these persistent wormers are not used these insects are relatively common.

Another casualty of these wormers was the handsome bird called the chough. Although changes in livestock management and the attention of egg collectors were probably both to blame, choughs became extinct in Cornwall in 1952. This was very inconvenient because the chough was the emblem of the county. Fortunately, the reintroduction of cattle to graze the cliff tops and the use of non persistent wormers led to their return in 2002. Alternative non persistent wormers

Tramlines for the crop sprayer are the chemical signature on our food and our wildlife

do, of course, exist and parasitic nematodes can be controlled by rotational grazing and by reduced stocking rates.

The effect of chemicals used in arable farming, in heavy applications of chemical fertiliser on grassland and through persistent wormers has been so enormous that it is salutary to ask the question, "How much land in your parish has managed to avoid these chemicals, and where are the plants, fungi and insects that remain unaffected?" I did a quick survey in our small parish of Winterbourne, which lies just to the north of Newbury, but which has within its boundaries 60 ha (150 acres) of ancient woodland and 70 ha (175 acres) of common land. This amounted to 15 per cent of the parish. When I looked at our neighbouring parish, the percentage of land unaffected by chemicals was barely 10 per cent. In the Midland shire counties of England, where woodland and common are both scarce, the land unaffected by chemicals may be considerably less.

However, the greatest challenge when we are attempting to restore wild flowers is the huge increase in fertility and its consequences. Fertility helps grass to grow and grass out competes the wild flowers. Fertility today comes in a sack marked nitrogen. You simply buy it and apply it with a fertiliser spreader. Before about 1900 fertility in the countryside was extremely hard to come by and there was only one source – animals. The manure heap was a feature of life.

On all but the lightest land cattle were kept in yards over winter and the precious manure was spread on hungry land to grow better hay or corn. Fertility limited food production, which was why most farms were mixed farms where grass was grown for three or four years and then ploughed up to produce the fertility to grow a corn crop.

This cheap and easy source of fertiliser has not only made the corn prairie a practical possibility but has also led to the profligate use of fertiliser, which has had a serious impact on a whole mass of plants including wild flowers. ICI perfected the manufacture of fertiliser in granular form so that it could be flung 5 or 6 metres (15-20 ft) either side of a tractor.

Where toxic wormers are not used, hornet robber flies are common

Before the advent of chemical fertiliser, fertility was hard to come by

The Wildlife Pyramid
A great mass of green plants, including wild flowers, at the base of the pyramid supports everything above

The farmer is able to drive over his fields with ease but if great care is not taken the fertiliser is flung into the hedge.

In the "gung-ho corn everywhere" days of the 1980s, it was not uncommon to hear fertiliser granules rattling on the car windscreen as you drove down country lanes in early summer. This was not only wasteful but it all helped tip the scales against wild flowers, which need to compete in order to survive. The common enemy is grass, which grows extremely well in the British climate with its (usually) abundant rainfall. Without careful management, grass will out-compete wild flowers on many of our soils but given a little fertiliser wild flowers will gradually decrease until most simply disappear.

The consequences of fertilisers go one stage further. When much of a farm is ploughed up, there will be fragments of wild-flower-rich hay meadow that survive. I met a farmer a few years ago who recounted his dilemma over this. Creeping thistle grows superbly if it is fed nitrogen, and it had spread from one of his arable fields to his old hay meadow, where it was increasing. What was he to do? He sprayed the hay meadow with a herbicide to control the thistles and that was the end of most of the wild flowers. Today, we have the invention of the weed wiper (see chapter 9), which enables us to control the thistle but save the wild flowers.

By an odd coincidence, a study was carried out in 1983 into the effects of this agricultural intensification a mere ten miles from where I live north of Newbury. The statistics are chilling:

Between 1947 and 1981	
Hedges	-42%
Hedges with standard trees	-46%
Visible footpaths and tracks	-54%
Ponds	-68%
Woodland	-25%
Arable	+38%
Permanent pasture	-79%

What must also be remembered is that those hedges and ponds that did survive only did so in a very degraded state, so the losses overall for wildlife were catastrophic.

It is a depressing story. The intensification and simplification of our agriculture with the demise of the mixed farm and the removal of much of our countryside, which was followed by a massive onslaught of pesticides and herbicides, has resulted in a huge loss of diversity. Many of the characteristic plants of our counties with their different genetic make-up have gone for good. But however depressing the story, there are lessons to be learnt if we want to make a success of the restoration that is now under way.

A diagram of the food chain shows species such as birds and spiders at the top of the pyramid. Below them is a whole mass of small mammals and insects, and at the bottom a wide range of plant species that support everything above. If these plants can be restored – and we now know that this is possible – many of the species that depend on them will return.

SUMMARY

Wild flowers were gradually diminishing as agriculture intensified, but when Britain joined the Common Market in 1973, corn prices quadrupled to £120/ton and every acre of land was ploughed to grow corn.

- Many farms were converted to arable, and remaining hay meadows and wild flower pastures were ploughed.
- Arable farms need chemicals for weed control and for fertiliser; most farmland is now fertilised, which tilts the scales against wild flowers.
- With the advent of effective chemicals, autumn sowing replaced spring sowing with the loss of winter stubbles, which were a major winter food source for wildlife.
- Persistent chemicals were introduced to worm cattle. This sterilised their dung, drastically reducing a whole range of insects.
- A Berkshire study of the loss of hedges (-42%), ponds (-68%), woods (-25%) and pasture (-79%) tallies with the subsequent loss of farmland birds (-50%).

above: The miraculous survival of some cowslips in a sheltered hollow

top: Horses mowing hay before the advent of tractors

pages 18-19: Red clover, which is particularly valuable for bumble-bees, originated from dormant seed at Hampshire Butterfly Conservation's Reserve at Magdalen Hill Down, Hampshire

CHAPTER 2

WILD FLOWERS – AN INTEGRAL PART OF RURAL LIFE

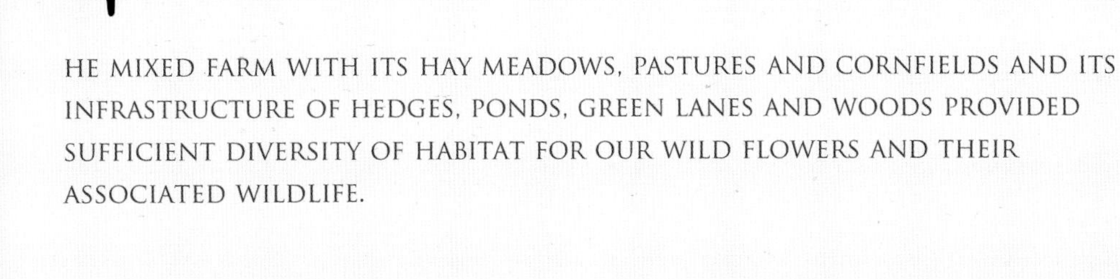

THE MIXED FARM WITH ITS HAY MEADOWS, PASTURES AND CORNFIELDS AND ITS INFRASTRUCTURE OF HEDGES, PONDS, GREEN LANES AND WOODS PROVIDED SUFFICIENT DIVERSITY OF HABITAT FOR OUR WILD FLOWERS AND THEIR ASSOCIATED WILDLIFE.

The heyday of this period dates from the enclosures of the eighteenth century, when hedged fields with mixed farming and its rotations replaced the low output of the open fields of the medieval period. Some of these management practices continue today, some do not. But it is essential for us to understand them if we are to make a success of restoring our wild flowers. It is a common perception that protection is the solution to a scarce or declining plant, but this can often only make matters worse. It is in fact consistent annual management that is the more important and, contrary to popular perceptions of the countryside, the richest wild flower habitats are those that are most intensively managed.

THE HAY MEADOW

The medieval farmers went to enormous efforts to find fertile, level valley bottoms where they could establish hay meadows, particularly in the populated Midland shire counties, where most of the land around the village had been cleared for the open fields.

Some of these hay meadows have miraculously survived, particularly in the Thames valley. They would have existed in most parishes, the larger ones where the population was relatively high, with numerous smaller ones to serve isolated farmsteads and hamlets.

It is a sobering thought for us today that a successful hay crop was the only means of feeding animals, particularly breeding stock, through the winter. Since there is so much to say about the former hay meadows, they will be discussed in detail in Chapter 3. However, as far as the farming context is concerned we can manage our hay meadows today in the same way as they were managed four hundred years ago, and with modern equipment the job is made a hundred times easier. There is no reason why the wild flower meadow should not make a comeback especially as new research suggests that the production from a field of mixed grasses and wild flowers can tap into deeper levels in the soil and achieve very reasonable levels of production.

Modern tractor mowing

When the crop is so close to the hedge wild flowers don't have a chance

HEDGES

The insidious degradation of our hedges may be coming to an end. New rules are now in place to help safeguard hedges by banning all spraying and cultivations for a minimum of 2 m (6 ft) from the centre of the hedge. This is a huge step forward. Almost all these hedges will be trimmed with the tractor mounted flail, which trims the hedge to a rectangular or A-shape and probably also cuts back the scrub and grass in the hedge bottom, thus giving the wild flowers sufficient light to survive.

Before the days of wire (introduced after the first World War) and the introduction of the flail it was all very different. The agricultural labour force was substantial and the winter months were spent hedging and ditching, with hedges being cut and laid in rotation every eight to twelve years.

During these years, as the hedge grew wider wild flowers and grass would have been shaded out for up to 1 m (3 ft) either side of the hedge but when it was cut, a metre of bare ground was available for the wild flowers to re-colonise. When we lay new hedges at our farm, we see this process at work and it gives us an opportunity to reinstate wild flowers along the hedge.

Although the cut and laid rotation is exactly as practised for hundreds of years, it is so labour intensive that we can only afford to do it very occasionally, but we aim to cut and lay a new hedge around ten years after planting and then rely on the flail for the rest of the time. The outlook for wild flowers along the hedge has improved, but arable cropping is still far too close for much to survive.

Another development which has come to the aid of the embattled hedge is the introduction of the 6 m (20 ft), or wider, field margin. These were originally placed against field boundaries because of the weeds which inevitably occur, particularly along hedges, and which then contaminate an otherwise weed-free crop. These margins started off by being grass only but increasingly they are seen as an opportunity to help wildlife, through the addition of wild flowers and where they can act as corridors for insects and animals from one part of a farm to another.

PONDS

There is a wonderful revival in ponds and lakes, thanks to the advent of the butyl liner, although

Village pond with trampled margin. In reality, trampling may have been so complete that only a few tough plants could have survived

Hurst village pond with attractive marginal vegetation and recreational seat

Sheep penned with hazel hurdles. Hurdle-making was a major rural industry that employed hundreds of families

the waterlogged grassland or wetland has yet to make much of a comeback. Paradoxically the very agents of the former destruction of our ponds are now working overtime to recreate them. I speak of the drivers of the JCBs (named after the company J.C.Bamford which makes them) and tracked excavators. In the 1980s you would try in vain to persuade any of these stalwarts to excavate a wetland feature with gently graded banks.

Drainage ditches had vertical sides, and if you wanted a pond you got vertical sides. Today you only have to mention the magic word "wildlife" and you will get gentle gradients, ledges for aquatic plants at various depths and much else besides.

It is hard to compare ponds or wetlands of 100 years ago with the situation of today. Wetlands were commonplace with a great variety of plants, whereas now they have been drained and turned into productive farmland. Although ponds have greatly decreased in numbers, the ones that survive are now managed for wildlife or recreation, whereas their use was strictly for animals with much trampling around their margins and in the case of village ponds very little in the way of marginal aquatic plants. Village ponds also had a procession of farm carts being pulled in and out during hot weather to help swell the wooden wheels and prevent the iron rims from falling off. Not only are ponds on the increase, but there is a realisation that the last 30 years of land drainage has greatly contributed to flooding downstream and that farmland is the ideal place to store water, so wetlands in some form will also be making a comeback.

WOODLAND

Coppice. Hazel coppice was an important type of woodland, ubiquitous in Britain with substantial areas on the higher chalk and limestone soils in Dorset, Hampshire and Wiltshire. In Tudor times when sheep and wool were the mainstay of England's wealth and foreign trade, the best land in each parish would have been enclosed for growing hazel coppice, such was its importance to the medieval economy, especially for producing hurdles. Right up to the arrival of wire at the beginning of the twentieth century, hurdles were essential for penning sheep.

There was the remarkable system in chalk downland areas of sheep being grazed on the downs with a shepherd in attendance and then being confined at night by hazel hurdles on lower ground where a spring crop was to be grown so that their dung could provide the fertility. These hazel coppices, which were cut every seven to eight years, provided the greatest diversity of woodland flowers. The secret once again lay in the regularity of the management.

After the First World War most hazel coppices became derelict. Anyone attempting to revive a hazel copse today has two problems. First, there is the time it takes to cut and lay the old hazel to get the stools (or clumps) back to the required density of two metres (6 ft) apart and this alone may take two coppice cycles or fourteen years. Second, there is the explosion of the deer population, muntjac and roe, which may mean the need for a deer fence. So the investment will be high but although there is almost no demand for the hazel hurdle, there is a growing demand for 1.5 m (5 ft) stakes for the hedging revival as well as the slender bindings woven along the top of the newly laid hedge between the stakes.

If demand for stakes and bindings is insufficient, it may be possible to grow hazel on a longer rotation for biomass production. Thus, the demand for hazel coppice looks reasonably bright even if the problems in restoring a derelict hazel coppice are formidable. There is also sweet chestnut coppice, largely restricted to Kent, Surrey and Sussex and cut on a ten to twelve year rotation for chestnut paling.

Other woodland. Coppice represents only a fraction of our woodland. What of the rest of it?

The hazel coppice provides the greatest diversity of wild flowers and other wildlife

With the revival of hedging, there is growing demand for stakes and bindings

It seems astonishing that most of the small woods in the south of England are derelict and it is only the larger estates that have the resources to keep some sort of management going. So how has this affected woodland wild flowers? Plants that thrive in shade are doing well in neglected woods. These include blue-bells and species such as early purple orchid, which like to seed onto a bare woodland floor with little or no competition.

For the majority of woodland plants it is an unsatisfactory state of affairs, and most have declined significantly. Primroses can survive in shade and in quite intense grass competition. Yellow archangel and wood anemone can grow in deep shade because they grow and flower before the leaves come out on the trees. Bugle, greater stitchwort, violets and sweet woodruff just manage to hang on along ride or woodland edges. Of course, if a tree blows down, the increased light stimulates all these plants to grow as well as others such as foxglove, herb robert and wood spurge, but after a few years, if there is no management, the clearing will fill up with brambles.

The cessation of hazel coppicing has had a drastic effect on dog violet. Its decrease has in turn had a very significant effect on a group of woodland butterflies, the Fritillaries, because they use it and nothing else to lay their eggs on. When the coppice industry was in full swing, there would be areas of coppice all over Britain where the hazel was cut every year. It was in the years immediately after cutting that the violets would be at their most lush and this provided the most suitable conditions for these butterflies. Come the demise of this industry with the arrival of wire, the violets declined

Yellow archangel can survive reasonably well without coppicing but will not flower unless the light is let in

massively. Today, the Pearl-bordered and Small Pearl-bordered Fritillaries have gone from being ubiquitous to being rarities and the High Brown Fritillary is restricted to the Lake District. It is only the Silver-washed Fritillary that remains reasonably common in the larger woods. Without a revival in coppicing it is hard to see the fortunes of these butterflies recovering. The Fritillary butterflies' reliance on dog violets emphasises just how long man has been coppicing in this country. Some of the ancient trackways in the Somerset levels date from around 3700 BC. They were made from managed hazel coppice!

ROADSIDE VERGES

Until after the Second World War a "lengthman" complete with scythe was employed to cut back the vegetation and look after the verge along every road and lane. It is hard to imagine the scene, but this was before the motor car had taken over our lives.

As a result, the verge with its wild flowers was cared for and the cut material was removed and stacked here and there as the opportunity arose. Today, many of our verges are pathetic reminders of past glories, having suffered the effects of fertiliser and herbicides from adjoining farmland, which has vastly increased the growth of grass or weeds such as hogweed, with now the added indignity of the roadside flail. This has been another disaster for our wild flowers, because the verges in many parts of the country contained the last vestiges of local genetic diversity. The reason for this disaster is that although the verges may be cut twice a year, the cut material is left as a thick mulch all over the verge.

The result is predictable: the suppression of most of the wild flowers and their destruction over a few years. Only the tall and the tough can survive this treatment: cow parsley at the beginning of the season and, at the end, meadow cranesbill, which also has a mighty root stock that enables it to force its way up through the mulch.

Some counties are beginning to organise the management of diverse verges by blowing the cut material into a hopper and removing it. But as things stand at present, the outlook for roadside verges on level ground is bleak.

top: Coppicing is essential if dog violets are to flourish

above: Wood anemones thrive in unmanaged woods

opposite: Greater stitchwort survives well along rides and the woodland edge

Cow parsley has a competitive advantage, since everything that flowers later gets mulched

Corn right up to the verge – no hope for wild flowers here. Whenever a hedge along the road is restored wild flowers have a chance

Nettles in the verge – even if a hedge exists nitrogen may seep through and cause dense nettle growth

The roadside flail is all we have to manage our verges

After a visit from the flail, a mulch this thick is usually used to suppress weeds

Meadow cranesbill is strong enough to grow through the mulch at the end of the season

Fortunately banks are less susceptible because much of the cut material falls or gets washed down the slope. There are, of course, exceptions where wild flowers flourish on thin soils where there is little fertiliser influence or where an enlightened organisation has sown the verges beside its entrance.

The above is a crude simplification of the key features that are necessary for wild flowers to be restored.

There is no obstacle to restoring the traditional hay meadow, as many farmers and landowners have demonstrated. Hay meadows without animals pose a different problem, which will be discussed in Chapter 8.

Hedgerow wild flowers, too, have a promising future, as do most of the pond aquatics. However, the wild flowers of the hazel copse face formidable problems because they require such an enormous investment in both time and money. And finally, the future of the roadside verge, which for most of us is entirely out of our control, looks extremely bleak.

However, before we set out to describe the opportunities and hazards involved in restoring our wild flowers, a more detailed description of the traditional hay meadow is essential because it has so many relevant lessons for us today.

SUMMARY

The mixed farm of the nineteenth and twentieth centuries maintained a wide range of habitats such as meadow, hedge, woodland, ponds and wet meadows, which suited many species of wild flowers and their associated wildlife. Can we create suitable conditions under the intensive, simplified conditions of today's farming? Fortunately the answer is largely in the affirmative.

- *The hay meadow:* modern equipment has transformed the hay-making process. The hay meadow could be in for a revival
- *Ponds*: these are being created in increasing numbers, but to be really useful for wildlife they need to be planted up with the appropriate species
- *Woodland*: although the restoration of hazel coppices may be only for the enthusiast, it is inconceivable that our woods will remain derelict with the opportunities for biomass production and other timber products. New woodland is not yet seen as an opportunity for introducing wild flowers
- *Roadside verges*: until improved methods of cutting our verges are adopted, any remaining wild flower diversity will continue to decline

opposite: Bank of wild flowers on the way down to Shalbourne village. Thank goodness for steep banks!

CHAPTER 3

THE TRADITIONAL HAY MEADOW

BEFORE WE SET ABOUT PLANNING A RESTORATION PROJECT, THERE IS ONE IMPORTANT SUBJECT TO EXPLORE IN MORE DETAIL, NAMELY, THE TRADITIONAL HAY MEADOW SYSTEM.

An understanding of this is crucial, because invariably hay meadows were located on fertile soils, which is very much the situation facing us today.

Making hay was a universal activity wherever animals were kept. Breeding stock, dairy cows, horses and oxen had to be kept through the lean winter months, particularly the period from January to March before grass growth resumed. The scale of this haymaking varied hugely, from tiny meadows and bits of common to large areas of 20 hectares (50 acres) or more such as North Meadow Cricklade in the Thames Valley, which has been in existence for over a thousand years.

The larger meadows were concentrated in the Midlands and in the south of England, where there were wide flood plains with alluvial soils. Many of them are recorded in the Domesday survey and originally provided the winter fodder for the oxen plough teams that worked the medieval open fields. When horses replaced the oxen and as agricultural efficiency increased, the number of horses on the land in the decades before the arrival of steam power must have been prodigious resulting in a huge demand for good hay.

Given the demand, it was necessary to mobilise considerable manpower to "make hay while the sun shines," which was as important then as it is today. When the horse-drawn mowers had done their work, whole communities would turn out to turn the hay and reduce the moisture in it until it was dry enough to cart away. Every bit of hay was raked up and a stream of carts took it to the farmyard where a hayrick was expertly built with the final finish being a thatched top to keep the weather out.

opposite top: Food production relied completely on horse power, and good quality hay was as important as diesel is for our tractors today

opposite below: Considerable manpower was needed to gather in the precious hay

pages 30-31: A hay meadow on the chalk in Wiltshire with abundant greater knapweed, wild carrot and scabious

KEY SPECIES OF THE HAY MEADOW

As will be discussed in more detail later, there are some forty species commonly found in a hay meadow, with an additional forty species being recorded less frequently. All these species have to flower and set seed before the cutting date in late July because species that flower later than this will not survive.

This does not pose a problem for the grasses, most of which tend to set seed before the end of July, but it is a problem for the wild flowers, which generally flower and set seed later than the grasses.

The list below is a useful guide as to how many species we should concentrate on. It is in many ways a remarkably short list considering that the average Flora contains 1,400 or more species. The main point is that this nucleus of a dozen grasses and a few more wild flowers can provide a basis for wildlife to re-colonise in a very significant way. In years to come important associations will no doubt be discovered between particular insects and additional wild flowers, but this list will take us a giant step forward.

Grasses

Cocksfoot (*Dactylis glomerata*)
Common bent (*Agrostis capillaris*)
Crested dogstail (*Cynosurus cristatus*)
False oat grass (*Arrenatherum elatius*)
Meadow foxtail (*Festuca pratensis*)
Perennial ryegrass (*Lolium perenne*)
Quaking grass (*Briza media*)
Red fescue (*Festuca rubra*)
Rough meadow grass (*Poa trivialis*)
Sweet vernal grass (*Anthoxanthum odoratum*)
Yellow oat grass (*Trisetum flavescens*)
Yorkshire fog (*Holcus lanatus*)

Bailing hay in Winterbourne Park. Mechanisation has transformed the task of bringing in the hay

Wild Flowers
Autumn hawkbit (*Leontodon autumnalis*)
Birdsfoot trefoil (*Lotus corniculatus*)
Common sorrel (*Rumex acetosa*)
Common vetch (*Vicia sativa*)
Cowslip (*Primula veris*)
Field scabious (*Knautia arvensis*)
Lady's bedstraw (*Galium verum*)
Lesser knapweed (*Centaurea nigra*)
Meadow buttercup (*Ranunculus acris*)
Ox-eye daisy (*Leucanthemum vulgare*)
Ribwort plantain (*Plantago lanceolata*)
Self heal (*Prunella vulgaris*)
Yarrow (*Achillea millefolium*)
Yellow rattle (*Rhinanthus minor*)

MANAGEMENT

The hay meadow had a number of other key features that need to be briefly examined.

The first of these is, of course, the management, with the key decisions being whether to graze the field or to shut it up for hay; and if it was to be shut up for hay, when to bring in the mower.

Since so many of our ancient meadows are now nature reserves, it is inevitable that they are cut for hay every year so that the public can appreciate their wild flowers. But for the small medieval grass farmer with a group of meadows, only a small proportion would have been cut, sufficient to provide for his stock through the winter, which may well have been longer and colder than the winters we are experiencing today. The remainder would have been grazed on rotation through the season. Management determines the proportion of grass to wild flowers, and particularly the amount of coarse grasses such as cocksfoot, Yorkshire fog and false oat grass. These grasses thrive on a lack of management and will dominate an area if it is left unmanaged. Usually, the more a field is grazed, the higher the proportion of fine-leaved grasses but the response of wild flowers is more complex. Lack of management will lead to coarse grasses crowding out the wild flowers, but continuous grazing, with no opportunity for the wild flowers to flower and set seed will also lead to a reduction of wild flowers. A balance has to be achieved.

But there is something even more important that, until recently, we have ignored or simply forgotten. All the old meadows were full of a grass parasite called yellow rattle, which had a major impact on their management. This is of particular interest to us today because, unlike the medieval farmers who were always looking for more grass, we are looking for less grass and more wild flowers.

Yellow rattle

Yellow rattle or hay rattle as it is often known, is a semi-parasitic wild flower annual of medium height. It can vary markedly between one part of the country and another. Being a parasite, it lives off the grasses in the meadow, thus reducing their yield. A small meadow at our farm used to produce two hundred and forty bales of hay, but when the yellow rattle was fully established this reduced to ninety bales, a reduction of over 60 per cent, but it has greatly benefited the wild flowers.

Farmers have always considered it a pest from the earliest times since it robbed them of precious grass and all grassland was probably full of it. In an old book on husbandry of 1756, there is a telling reference to the danger of yellow rattle. Today's farmers do not like it either, but since it is an annual, preventing it setting seed will eliminate it from a meadow in a few years. It also happens to be very palatable, so if cattle are let into a field, it is the first thing they will eat. Since it is so effective at

With correct management, yellow rattle will effectively control grass growth

reducing grass growth, we can turn this to our advantage to benefit wild flowers, so long as we can use the methods of the traditional hay meadow system.

Minerals and animal health

This is another area where we have managed to lose a great store of knowledge that would have been second nature to the medieval farmer.

Dairymen will describe their animals going into a new field of perennial ryegrass and immediately hunting round the edges of the field to pick at the wild flowers in the hedges. This is because many of the commoner wild flowers have deep roots or tap roots that grow into lower layers of the soil profile, thus making available additional minerals through their leaves. Such plants include birdsfoot trefoil, lesser knapweed, field scabious, common sorrel and yarrow. The minerals have a significant impact on animal health and on the flavour of the subsequent meat, but the actual detail of the contribution made by particular species seems to have been lost.

The cutting date

The date for cutting a meadow is always critical. Meadows always used to be cut around

Fine leaved grasses such as the fescues (left) usually indicate that wild flowers can be established. Coarse leaved grasses such as cocksfoot (right) indicate derelict sites or fertile soils where establishment will be more difficult

25th July, but there would be huge variations depending on whether there was a spell of dry or wet weather. The cutting date in medieval times was more important than today because if you got it badly wrong and upset the balance between grasses and wild flowers, it took much longer to put it right. At least today, if the species in the meadow become unbalanced, there are mechanical toppers and mowers that can assist at relatively low cost. So what is this balance? There was no way of keeping the wild flowers out of the grass before the days of herbicides, but the nutritional value of grass is at its peak in May, earlier than the peak flowering time of the wild flowers. These had to be able to flower and set seed if they were to survive. So it was very much a compromise between grass and wild flowers, one providing the bulk of the forage and nutrition, the other the key minerals.

If the cutting date got delayed because of poor weather, the quality of the grass would deteriorate. Delayed cutting for the occasional year has little effect on the balance of a meadow but if delays occur every year, the balance is affected.

The first thing that happens is that the stronger wild flowers such as the knapweeds get much bigger and begin to crowd out the weaker species such as lady's bedstraw. I know of several new meadows where there is almost a monoculture of knapweed, brought about by a combination of the owner's wish not to cut the knapweed before flowering is complete because of the abundance of butterflies feeding on it and the difficulty of getting someone to cut it because of the dearth of farm staff or contractors. An added factor is that new meadows are less stable than well established meadows and individual species can fluctuate greatly.

We carried out an interesting but unintentional experiment at our farm where a field was reseeded with a grass / wild flower seed mix. When it was fenced, a strip was left outside the fence and consequently unmanaged. After ten years, instead of fifteen species of wild flowers there were three: knapweed, ox-eye daisy, and common St John's wort.

The other factor is the process of natural selection that operates over much longer periods with regard to the cutting date. There will be a selective pressure on plants of all species that favours those that have seeded by

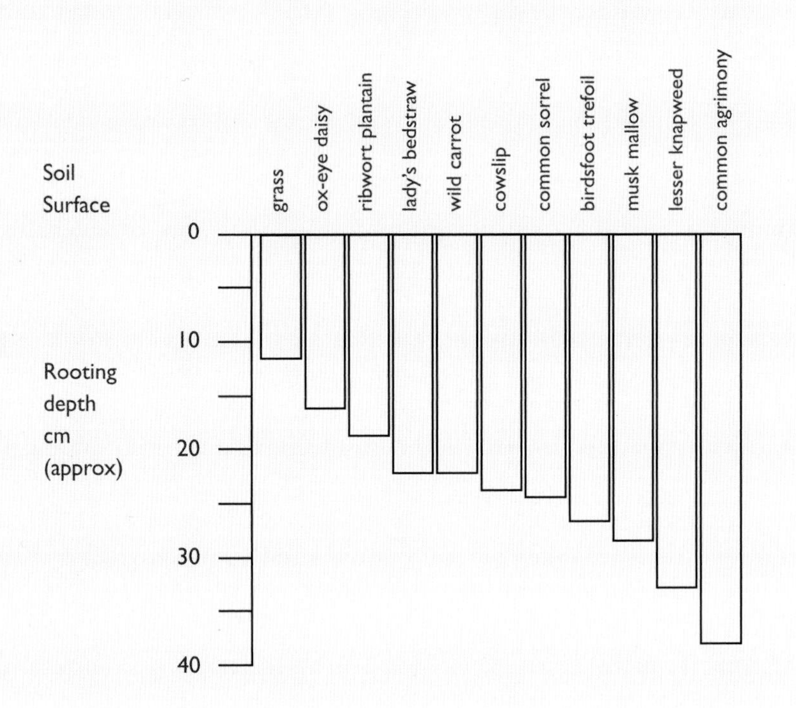

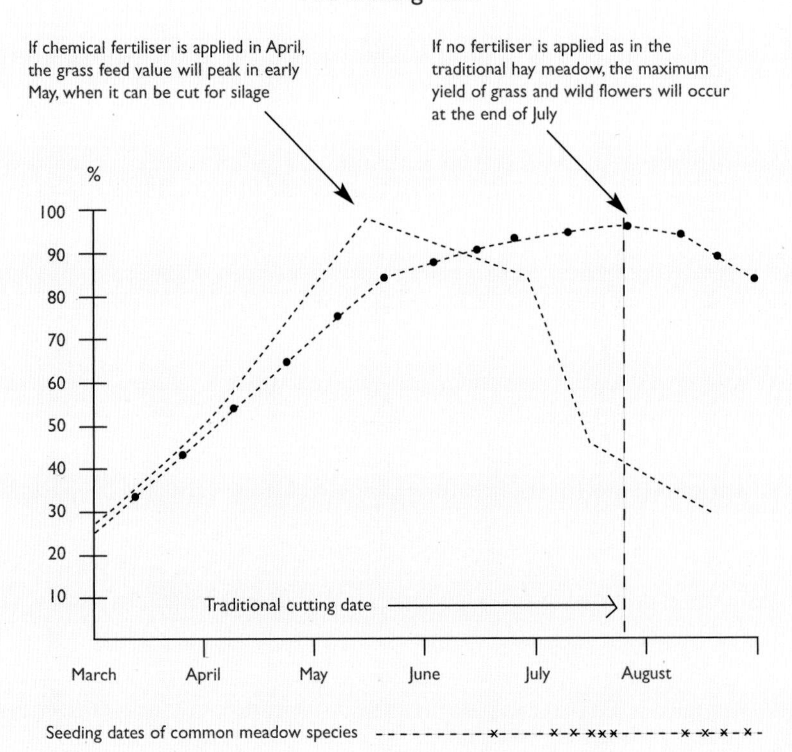

opposite: If cutting is continuously delayed beyond the traditional cutting date of 25th July, a single species may begin to dominate the meadow

25th July. Those that seed earlier will not be affected, but those that might seed later will not contribute seed unless there are a few plants with early seeds. Wild flowers such as agrimony and meadow cranesbill, which often seed too late for the July cutting date, may well be found in a hay meadow if early flowering plants have been selected over the centuries.

Grazing the aftermath

This was the last part of the traditional hay meadow process. Some six to eight weeks after cutting, the autumn re-growth would have produced 15cm (6 inches) of lush grass to graze in late September / early October. This grazing put off the time when the precious stored hay had to be fed but the removal of this growth, which was mostly grass, left the field ready for the next season's growth.

This autumn grazing also did something else very important. The grazing animal opened up the sward by pulling at it and eating it and then, by walking about trod in the seed shed from the year's hay crop. This resulted in bare earth having newly shed seed pressed into it. In the case of yellow rattle, there is a further factor here in that its seed is a lightweight disc, very thin but as wide as 4mm (one fifth of an inch), which can easily get caught up in dead vegetation rather than end up flat on the soil surface where it can germinate.

Grazing removes the autumn re-growth, leaving patches of bare earth and a complete absence of cut material that would otherwise prevent seeds coming into contact with the bare soil. Treading in the newly shed seed from the year's hay cut ensures that it gets good contact with bare soil and germinates successfully.

This system has been tried and tested over the centuries. What is proving more difficult is our attempt to modify it using the rotary mower without any grazing animals, of which more in a later chapter.

Where did the manure go?

The final aspect of the traditional hay meadow was the need to clear out the shed where the farmer's horses or other animals spent the winter eating the carefully stored hay. The manure spent the summer quietly rotting down in the shed, but had to be cleared out before the animals came in again, but not before the fields it was going to be spread on had been grazed down as tight as possible. These would be the fields to be shut up for the next year's hay cut. There was much co-ordination involved. So with the conventional wisdom stating that fertility was a bad thing, remember that the old hay meadows were fertilised with manure to help the grass growth, although by today's standards the amount of manure was fairly light.

RESTORATION

Restoration is now essential to safeguard threatened species. When all the hay meadows disappeared, many people felt that somehow it was better to wait for colonisation of new meadows to occur naturally rather than accelerate the whole process by adding new seed. Several developments have changed this view.

The status of certain species such as butterflies has become so precarious that we simply cannot wait for colonisation to take place over a long period. The risk of losing species is too great. This was tested at Butterfly Conservation's reserve at Magdalen Hill Down near Winchester where a 12-hectare (30-acre) extension, recently increased by a further 22 hectares (55 acres) was acquired next to the existing reserve and seeded with local seed. Within ten years, thirty-one species had been recorded on the 12-hectare extension. What has been truly remarkable is that the additional 22 hectares, seeded in 2004 had been colonised by 26 species two years later! A great deal has now been learned about using wild flower seed and there is much more confidence in the whole concept.

We have also learned that re-colonisation by perennials on arable land is not usually going to happen. This was shown to be the case when set-a-side was introduced in 1993. It was anticipated that the countryside would burst into flower from all the dormant seed in the topsoil. Sadly, nothing of the sort happened. Annuals such as poppies and speedwells were there, of course, in profusion as were biennials such as great mullein. But perennials, hardly one.

We found this on our own small farm. Arable crops had wiped out all the perennials, and if you are a perennial there is less need to build seed dormancy into your genetic make-up because, unlike an annual, you are going to live quite a long time. There has quite rightly been considerable doom and gloom about the hay meadow situation with both the statutory and voluntary wildlife agencies repeating endlessly that we have lost 93 per cent of our old meadows, so it is clear that we need to save those that survive. Yet it makes me wonder whether this negative attitude does anything to encourage the restoration process, which of course is the only way to repair the damage of the last thirty years. And there is a further problem with the restoration process in that endless arguments are still produced about the high incidence of phosphate and the way that this makes the establishment of wild flowers difficult if not impossible. Surely, where this is the case, the answer is to use yellow rattle.

It also appears that many people have never visited some of our old meadows, which all have high populations of yellow rattle. A read through old husbandry books confirms its universal presence, which transforms the wild flower situation where conditions are fertile. There are, it is true, the two constraints that a hay cut has to be taken at least one year in three to enable this annual to set seed and replenish itself, and the autumn re-growth needs to be grazed. But these are not onerous conditions. When I was learning about the possibilities of yellow rattle in the late 1980s, I experienced the problem of the lack of it and then the astonishing transformation of the meadow when it was added

In Northamptonshire, a 1.2 h (3 acre) meadow of red fescue and timothy had resisted all attempts to introduce wild flowers until yellow rattle was established, and in a fertile meadow in Berkshire, the grass was swamping the wild flowers until yellow rattle was added. From then on, we have used yellow rattle over several thousand acres, particularly on ex-arable land on the chalk, and it has been highly effective.

There is now enough evidence that we can restore wild flowers on almost any soil, however fertile. There is yet another belief that wild flowers, once established, deteriorate over the years but this ignores the essential role of

opposite: Seeding thirty acres of arable land at Magdalen Hill Down, Hampshire, for butterfly conservation resulted in the establishment of 31 butterfly species in 10 years, many in abundance

management, which is discussed in Chapter 8. It arises partly from the perception that the natural world is a result of non-intervention, and that management is somehow not quite right or appropriate. This, of course, flies in the face of all experience, which sadly bears out the fact that hay meadows, like many of our most cherished habitats, are the product of intensive management. The only way to maintain them or to re-create them is to continue this intensive management.

SUMMARY

The hay meadow provided the energy for farm horse power before the days of the tractor.

- Although the average Flora lists 1400 species, there about 25 key grasses and wild flowers which can help restore wildlife by providing nectar for insects.
- Management of hay meadows is complex and yellow rattle is an essential part of this.
- Many of our meadow wild flowers have deeper roots than grasses and can tap into lower levels of the soil profile and recycle minerals from these lower levels to farm livestock.
- The cutting date of 25th July is a compromise between the peak nutritional value of grass (mid May) and the seeding time of wild flowers (late July to mid August).
- Grazing the aftermath (re-growth) of a hay meadow has the added effect of treading in the seed shed from the year's hay cut, which is essential in the case of yellow rattle.
- Restoration of hay meadows through re-seeding is now vital if we are to halt the decline of insects in general and butterflies in particular. Excellent results can be achieved within ten years.

Cornfield annuals such as poppies are adapted to producing huge amounts of seed that keeps well in the soil

CHAPTER 4
PLANNING FOR BIO-DIVERSITY

AN ENTIRELY NEW MIND SET ABOUT THE COUNTRYSIDE IS IN THE PROCESS OF BEING CREATED. UNTIL ONLY A FEW YEARS AGO, ALL LAND MANAGEMENT (EXCEPT OF COURSE FOR NATURE RESERVES, SITES OF SPECIAL SCIENTIFIC INTEREST AND SIMILAR) WAS PLANNED FOR MAXIMUM PRODUCTION OF ONE OR TWO SPECIES, SUCH AS CORN OR GRASS AND CLOVER. THE MINISTRY OF AGRICULTURE WAS THERE TO GROW FOOD.

above: Species at the top of the food chain, such as this sparrow hawk, are not popular when they raid the bird table, but their presence indicates reasonable health further down the food chain

opposite: Worker bumble-bee (*Bombus pascuorum*) on birdsfoot trefoil. Note the pollen load on the tail

pages 42-43: Bumble-bee (*Bombus* sp.) on greater knapweed. Knapweeds have high quality nectar, which is ideal for bumble-bees

This Ministry has now been replaced by a new creation called DEFRA, the Department of Food and Rural Affairs, which is meant to be interested in the whole rural economy.

Its new Environmental Stewardship Programme is a variation on the old Countryside Stewardship scheme, but has now been offered in a simpler form to the farming community as a whole, with a discretionary scheme being available for farms with special environmental interest. The main point of all this is that if you are digging ponds, restoring wild flowers or generally diversifying the countryside, you are likely to get a sympathetic hearing, whereas in the days of maximum production, you were simply considered to be beyond the pale! If you are going to try your hand at this, and that is what this book is all about, some serious planning is required.

WHAT ARE YOUR OBJECTIVES?

You may be only interested in birds or wild flowers or bats, but I am going to assume that you want all the wildlife you can get your hands on. This makes everything much easier because even though we talk about our main farmland habitats of arable, meadow, hedge, pond and woodland as if they were self contained, they are all interrelated because of the significant number of species that require more than one habitat either to complete their life cycle or to survive through the year. Chaffinches may nest in hedges but spilt grain from winter stubbles helps to sustain them through the winter. The Brimstone butterfly needs the nectar of the hay meadow or any other wild flower it can find, but actually lays its eggs on purging buckthorn, one of our native hedge plants. Dragonflies, one of the exciting species to colonise any new pond also need the hay meadow because so much of their prey will be attracted to the nectar of the meadow flowers. Kingfishers, birds of stream and river, also come and prey on the dragonflies in the pond. And if there is a dry summer, birds, mammals and insects all come to the pond to drink.

Species of the hedge, such as the yellow-hammer are hugely reliant on any wild flowers along the hedge because they attract impor tant insects for their young. Barn owls hunt over rough bits of grassland, hay meadows and hedges with rough grassland at their foot because this is where they find their prey, the small field mice and voles. And the blackbirds, song thrushes and robins that find their nest sites in woodland, invariably feed in the meadow. All these habitats are inter-related, so there is a strong argument to provide all of them, with the hay meadow as a central feature because of the huge supply of nectar it provides. This multi-habitat approach will immediately begin to put right one of the biggest casualties of the countryside destruction, which was the elimination of groups of different habitats all quite close to one other, and their replacement by an arable monoculture.

Year round survival

You may have noticed that I included in my list of habitats the dreaded word "arable". Although much of England may have been

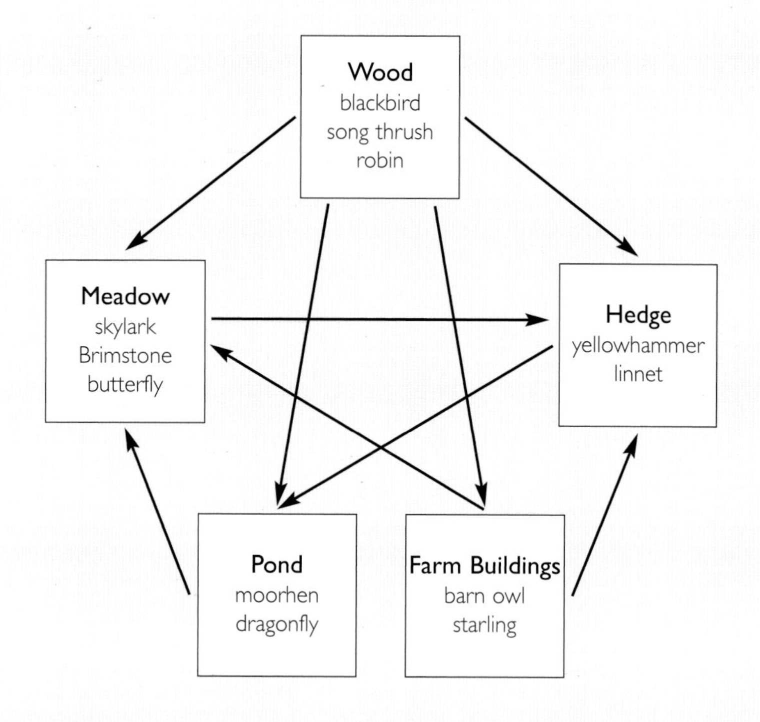

Habitat	Trees/ shrubs	Wild flowers	Birds	Other species butt'flies/drag'flies	Total
Meadow		25	6	20	51
Hedge	17	12	12	12	53
Pond	-	32	6	7	45
Woodland	18	25	13	6	62

Habitat	sq.m	%	Time to create
Meadow	2900	73	5 years
Hedge (perimeter 250m)	500	12	8 years
Pond (14m x 14m)	200	5	1 year
Wood	400	10	20 years
	4000	100	

turned into a corn prairie, the arable element is essential, particularly in winter, when food for wildlife can be in very short supply. Even with the efficiencies of the modern combine, winter stubbles, if they can be left through the winter, provide vital food for many birds such as partridges, finches and buntings. Spring crops were so important because autumn sowings removed all this food. Even if spring crops are impracticable, there is the option of providing small areas of wild bird food, which will be a lifeline for hundreds of birds through the winter. Provision for scarce farmland birds needs to be made all through the year, not just in summer, which is why planting magnificent hedges for yellowhammers is not the whole answer (although it helps greatly) unless winter food is also provided.

Habitats and species diversity

With many of our farmland species using more than one habitat, there is a strong argument for restoring meadow, hedge, pond and woodland where possible. But there are other reasons that suggest that going for a meadow alone may not be the wisest course of action.

Creating a meadow is much more challenging than creating other habitats. On thin, unfertile chalk, sands and gravels, wild flowers will grow well without too much competition from grasses, but on the great majority of our soils, which are fertile, grass competition is intense and the management needs to be of a high order to counter this competition. In addition, with the virtual disappearance of the agricultural labour force, providing any form of management is becoming increasingly difficult.

Habitats of hedge, pond and woodland may be peripheral to the meadow in terms of land use but they still score extremely well in terms of species diversity. In addition, if these other habitats are included, there will be the opportunity to provide nectar over a much longer period.

The top table above compares trees/shrubs, wild flowers, birds, butterflies and dragonflies but ignores all other species.

These statistics rely heavily on experiences at Carvers Hill Farm over the last twenty-five years.

The hedge habitat is a hybrid habitat between meadow and woodland. Although trees/shrubs score well, wild flowers score only 12, because they have minimal management. Species such as cow parsley, hogweed and similar, which are probably present are not included.

This table emphasises the point that although the creation of a meadow may be a key objective, it may not be a bad idea to create the peripheral habitats because they complement the meadow and have a very beneficial effect on species diversity. It is also significant that these peripheral habitats are generally easier to create.

There is a final point about land use. Hedges and ponds take up a very small area and woodland / shade can be on a very modest scale but still effective in creating the right conditions for many woodland plants. The bottom table assumes an area of 4000 sq.m (1 acre), with a hay meadow (73%), boundary hedge all round (12%) and a pond (5%), with woodland (10%). The three additional habitats are unlikely to ever take up more than a quarter of the area.

Having attempted to sort out our objectives, we can now consider taking action.

GROUNDWORKS

The aim here is to put into reverse everything that has happened in the last fifty years during which time the countryside was turned into a perfectly drained billiard table so as to achieve optimum conditions to

Horseshoe vetch

Horseshoe vetch seed, from which the plant takes its name

Seeds germinate reluctantly and seedlings establish only occasionally in the wild

Bloody-nosed beetle (*Timarcha tenebricosa*), which is often to be seen crawling lazily along in the grass

pages 48-49: A meadow with plenty of birdsfoot trefoil should have plenty of Common Blue butterflies. No birdsfoot trefoil, no Common Blues

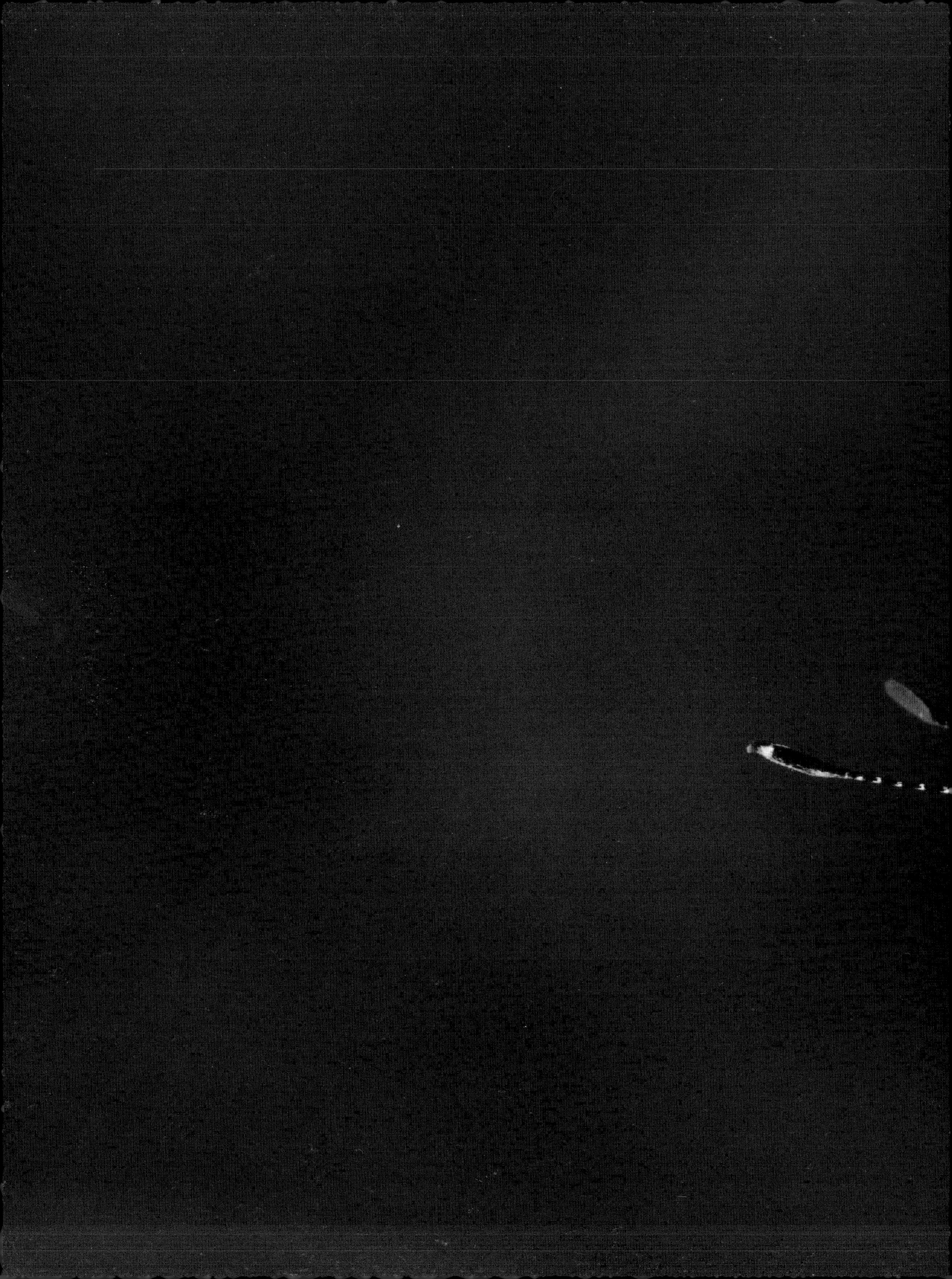

produce maximum crop yields. Depressions and hollows were filled in, banks levelled, wet spots drained and streams straightened or canalised. Now is the time to think the unthinkable and create diverse conditions for maximum diversity of wild flowers within the habitat that you are designing. This will lead to maximum bio-diversity as it will attract diverse wildlife. Although a session with an excavator is not essential, it will achieve conditions that will allow more diversity.

Banks and ditches

These tend to occur on the boundaries of properties, so it is logical to discuss them first. You cannot have a bank without a ditch. How steep should the bank be and how deep the ditch? It is vital to consider the future management. Banks too steep to be mown by a tractor are of little consequence if they are only, say, 25 m (80 ft) long. But what if they are 250 m long? Soil type and aspect are also of great importance:

Clay. Clay is an unforgiving material, suitable for gentle slopes. Banks 1 m (3 ft) in height should be 3 m (9 ft) wide, with a ditch to match. A hedge can be planted on the bank if required; the gradient can still be tackled by a tractor and topper but if the bank is too steep for a tractor, rank grass will soon invade and you will be grappling with scrub in no time. If there is a steady slope along the ditch, water will lie in the bottom and small bunds can be created at 25m (80ft) intervals to retain the water and prevent it draining away down the slope. A nice list of wet meadow plants will thrive there.

Chalk or Cotswold brash. This is completely different from clay. Water never lies in the ditch, and a steep slope of chalk subsoil requires little management since there will be insufficient fertility for grasses to dominate. Wonderful opportunities occur for special chalk wild flowers, such as wild thyme, common rockrose, harebell and horseshoe vetch. The latter is incredibly important as it is the food plant of both the Chalkhill Blue and the Adonis Blue butterflies, both greatly reduced in recent years. Gradients can be steeper, without too many worries about not being able to manage with a tractor. A south- or west-facing slope might be the optimum, but other aspects give many opportunities.

Banks and ditches do not have to be on boundaries. They can be anywhere so long as they do not make management too complicated. On clay, a flat field could be greatly enlivened by a wide shallow ditch across it, like a greatly exaggerated section of ridge and furrow with gradients gentle enough for a tractor and mower. If it fills up in winter, it will attract wintering snipe and who knows what else. On chalk, it is best to take an existing slope and emphasise it by increasing the steepness and removing the soil from the slope to create bare chalk for the special chalk wild flowers. Other species will spot an opportunity. Some mining bees moved into a small chalk cliff that we created on our farm.

Scrapes or temporary water bodies

The definition of a scrape is a gentle depression, say 1m (3 ft) in depth and anything from 10m (30 ft) x 10m (30 ft) upwards in area with an irregular edge or shoreline.

Scrapes are usually associated with wetlands, being permanent features where there is a high water table. Wading birds can feed either all year round or on migration. But there is untold scope for scrapes to be created well above the water table where there is a small amount of clay to hold water in the winter months. Our experience is that quite apart from attracting mallard, teal, lapwing and wagtails, snipe will call in and even a migratory green sandpiper, if you happen to be on a migration route.

The key feature of these scrapes is that they should be in permanent grassland that can be well grazed in the autumn so that there is no dense vegetation to hide the shoreline. As the water levels ebb and flow, the shoreline with its short grass will always be suitable for wading birds. There are also some very interesting insect dynamics, particularly in spring. As the shoreline retreats in warm weather, insects will hatch providing a continuous supply of food.

Permanent water bodies are ideal for many species, but their shoreline develops into a mass of vegetation very quickly, which is fine if you happen to be a moorhen but no good at all if you are a redshank. The key to increasing bio-diversity is to provide varied conditions.

right: Migrant hawker (*Aeshna mixta*) found in ponds, lakes and ditches

far right: Common green grasshopper (*Omocestus viridulus*). Jumps and flies actively from June to October

opposite: Fluctuating water levels expose muddy shorelines where wading birds can feed, but these areas must be grazed otherwise they become a wildlife jungle of little value

Ponds and permanent water bodies

Ponds are jewels of diversity with all the year round conditions for wetland wild flowers (which most of us never see today), their associated insects and most exciting of all, the inhabitants of the aquatic environment, amphibians such as frogs and newts, snails and dragonflies. All you need is a JCB or tracked excavator and a butyl liner. But do not forget planning permission if you live in an area where it is necessary. The possibilities are extraordinary and the pond can be up and running inside a year.

There will be more creative opportunities for doing something interesting with the spoil that comes out of the excavated area, such as creating a bank to protect the north or east side of the pond, or a bank to screen the neighbours. This will be discussed further in Chapter 12.

Topsoil

There is a lively debate about whether topsoil needs to be removed to help the meadow creation process. Some recent books on meadow creation show illustrations of JCBs hard at work removing it. The problem with topsoil is that it is the top section of the soil profile where growth takes place. If you are going for maximum growth, you want this part of the soil to be fertile, but if it is fertile, wild flowers come off second best to the grasses, which grow vigorously and then collapse onto the wild flowers creating a shambles in no time.

So is there a case for topsoil removal? There is, but only on certain soils.

Considering two of our main soil types, topsoil removal is fine on chalk or limestone but unwise on clay. On chalk or limestone wild flowers will be slow to establish but grass competition will be minimal. Two areas of 4000 sq.m (1 acre) each have been created at Magdalen Hill Down near Winchester in order to establish horseshoe vetch for the Chalkhill Blue butterfly. On another site a varied wild flower mix has been created.

Thanks to the topsoil removal both these sites have been successful. But if your soil is clay, beware! Removal of the precious topsoil removes all that lies between you and the clay, which is like concrete in summer and like plasticine in winter. If you are on clay, use the parasitic yellow rattle instead.

Another approach to growing wild flowers on fertile soils is to go for "nutrient stripping," which is planting a nutrient-hungry crop in the years before the wild flowers are sown. There is little evidence to suggest that nutrient levels are significantly reduced and there is then the additional problem of disposing of an unfertilised crop.

SUMMARY

- Many species require more than one habitat to complete their life cycle, so there is a great need for a mix of habitats.
- Mixed farming creates diversity of habitat, with winter stubbles being particularly valuable as a source of winter food for farmland birds.
- Creating hedge, pond and trees as well as a meadow will provide nectar from March to September and increase species diversity by as much as four times with only a small increase in land use.
- Creating banks, ditches, hollows and scrapes will reverse some of the levelling and drainage of recent years and increase habitat diversity.
- Clay and chalk soils require very different treatment if banks and cliffs are to be created.
- Temporary water bodies with grazed shorelines are as valuable as ponds with margins of aquatic plants.
- Topsoil removal is only practicable on small areas but can help maintain wildflower diversity, particularly on chalk.

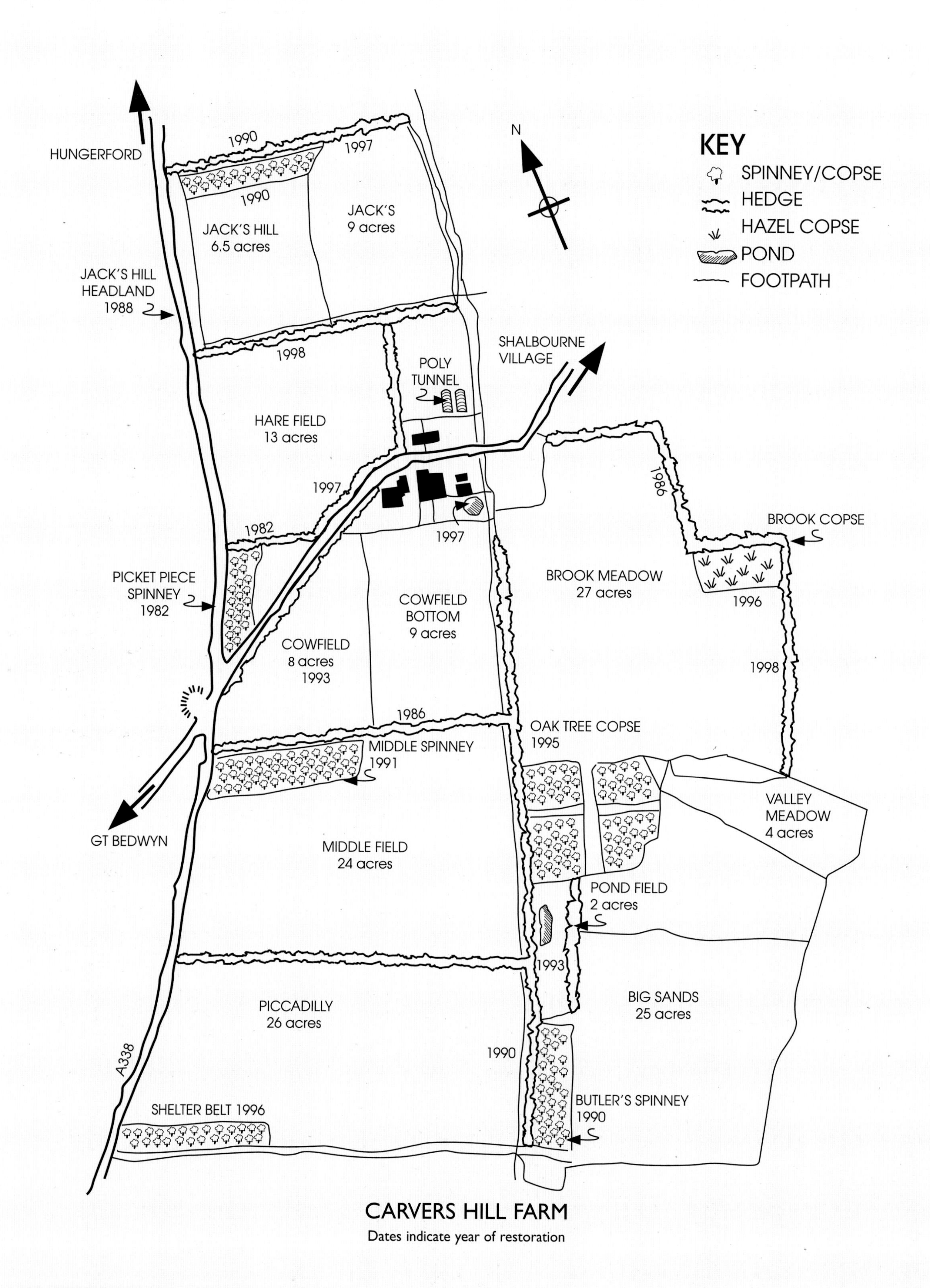

CARVERS HILL FARM

Dates indicate year of restoration

CHAPTER 5
CARVERS HILL FARM – THE STORY OF THE FARM

IT IS NOW JUST ABOUT TWENTY-FIVE YEARS SINCE I WAS SITTING AT THE BACK OF AN AUCTION ROOM IN A HOTEL IN MARLBOROUGH, WILTSHIRE BIDDING FOR OUR FARM.

I knew nothing about farming and not a great deal more about restoring farm habitats. I was convinced that there was a contribution to be made in terms of helping others to restore the countryside, especially those areas where there was no track record such as restoring wetlands and wild flowers. In 1980, everyone was ploughing up every bit of land they could lay their hands on to take advantage of the sky-high corn prices. So why was I so convinced that the pendulum was going to swing?

THE DREAM

I had been working for a small dynamic charity called the British Trust for Conservation Volunteers, which trained young people to carry out conservation work, such as restoring silted up ponds, laying hedges and repairing dry-stone walls. In the 1960s and 1970s a volunteering spirit was emerging where young people were willing to pay a small sum of money to go to remote and beautiful parts of the country and work flat out for a week or a weekend to put right some of the neglect of recent years, at a time when our farmers could think of nothing but larger heaps of corn. My job was to extend the number of regional teams and to co-ordinate their activities. I always travelled by rail. As I looked out of the train window, I witnessed arguably one of the greatest periods of destruction ever to be visited on our countryside. Piles of burning hedges and small and large copses were commonplace. This was state-sponsored destruction, all in the cause of the perfectly laudable aim of, as the White Paper described it, "Producing more food from our own resources." Like other subsidised schemes, it got completely out of hand. Of course the madness would end one day, because we love our countryside, but it was extraordinary working for an organisation that inspired young people to do so much good in the countryside at the same time as the farming industry and the Ministry of Agriculture were systematically destroying it. As I watched all this from my warm railway carriage, I dreamed one day of getting hold of a small farm and helping others to restore the countryside. Of course, it was only a dream!

BUYING A FARM

Within five years, a family trust had been sold and I had received just enough money to buy a small farm, but as the price of corn soared so did the price of land, and the amount of land I would be able to buy steadily shrank. Then Carvers Hill Farm came on the market with its 71 hectares (175 acres) and an interesting range of soil types. I acquired it in December 1980. It was, of course, far too small to be a viable unit on its own.

A share-farming deal with a cousin in a nearby village took care of the first three years but even at that early stage my complete lack of understanding of the realities of farming was revealed by the drainage scheme installed to grow corn on part of the farm that lay wet in winter. Remember that my aim was to restore trees and hedges. Imagine my horror on realising that when you put in a drainage scheme and there are hedges in the way, the hedges have to go. It is as simple as that. An inauspicious way to start restoring the countryside!

With the support of Kennet District Council, the first hedge went in as well as the first trees but the learning curve was steep. The share-farming scheme meant that there was no one on the farm except when a huge tractor came over the hill and ploughed and harrowed so

above: Why should anyone want to buy an unviable farm of 71 h (175 acres)?

opposite: The farm today. In 1980 there were no woods or ponds, and the land south of the farm buildings comprised two featureless cornfields of 29 and 27 h (72 and 67 acres)

pages 52-53: Undreamt of wild flower diversity can be established on poorer soils such as chalk here at Shalbourne

that the corn could be sown. There were occasional visits from the sprayer, which roared up and down the tramlines dispensing chemicals and then nothing much happened until the combine arrived and then the whole cycle began again. Because the farm was so quiet, the hares demolished most of the new hedge and the roe deer waited patiently until the new trees emerged from their 1.2 m (4 ft) tree tubes in Oak Tree Copse and then mercilessly pruned them.

MY FARMING HITS THE BUFFERS

The really risky part of the farming story began in 1983. I was determined to farm myself, understand about seedbeds and cropping and what the man on the tractor experienced as he worked the huge new cornfields. We were still an all-corn farm, and with part-time help with the ploughing and the support of the Stone family from Great Bedwyn, who carried out all the complicated spraying, we collected the necessary equipment to do the rest: tractor, second-hand plough, harrows, power harrow, combine, corn trailer, etc. But we had after all a tiny, unviable farm, only propped up by the ridiculously high price of corn. All went reasonably well until the price of corn began to dip in 1987. We had a really tough harvest, scooping up sodden wheat and barley and the repair bill for the combine alone went into four figures. It was not long before the bank manager started asking questions about the mounting farming losses. I had to face up to the fact that if I continued losing money at this rate, I would have to sell the farm, so a share farming arrangement was set up with the Stone family and I was free to start alternative enterprises, the key one being wild flower seed crops. During these five years progress was made on restoring part of the farm landscape. Two important hedges were planted in 1988 to break up a 29-hectare (72-acre) field and make three fields – Piccadilly, Middle Field and Cowfield – and we created the pond in Pond Field, which was extended a few years later.

INTO WILD FLOWERS

I had started growing contract wild flower seed crops on a small scale on ex-arable land for one of the large seed houses a few years earlier, and in May 1988 I sowed my first field margin of a quarter of an acre with one of their proprietary wild flower seed mixes on a part of the farm where the soils were thin chalk and where the rabbits usually ate any crop that attempted to grow. There was nothing to lose. To my surprise, the wild flowers all grew exceedingly well despite the fact that according to some of the statutory conservation agencies, wild flowers were not meant to grow on fertile farmland. This clearly needed further investigation.

I sowed the other half of the field margin the following autumn. This time things did not go to plan. The seed was sown late in October. November produced low temperatures and it was far too cold for anything to germinate. The following spring was cold and dry and again nothing germinated. I was all set to plough it up and forget the whole idea, but I never got round to it and to my huge surprise everything then germinated perfectly the following autumn. This taught me an important lesson. When wild flower seed is a bit slow to germinate, do not make hasty judgements.

THE STORY OF COWFIELD

In 1993 we signed up to the new Countryside Stewardship scheme, which for the first time paid farmers to manage the countryside rather than simply grow more food. We set up our first field-scale trials in 3.25 h (8 acre) Cowfield on a chalk bank. Little did I realise how important this trial was going to be.

I wanted to test two theories: that wild flowers would grow perfectly well after an arable crop; and that much lower rates of wild flower seed were feasible in the farming context.

In addition, I was determined to have the field monitored so that I could demonstrate the results. I was fortunate in persuading Dr Chris Smith, author of the classic work on *The Ecology of the Chalk*, to record wild flower establishment and to write up his findings. Effective monitoring has been sadly lacking with so many of our countryside initiatives.

The paradox of wild flowers after corn is that getting the crop established is remarkably easy because the only competition will come from annuals and the high residual fertility makes the wild flowers grow exceedingly well. Most perennials on arable land have been eradicated by years of herbicides.

All that needs to be done is to top the growth of annuals two or three times so that the newly sown seed is not shaded out. However care has to be taken that grass growth in later years does not swamp the wild flowers.

A friend rescued the seed barrow from an antique dealer on the grounds that I was actually going to use it! A good man could sow twenty acres a day with one of these

Cowfield was sown in May after a corn crop the previous year. Because of the high fertility, we cut it for hay for three or four years before the fertility began to drop. This kept the grass reasonably under control. Another theory that also needed to be tested concerned the seed rate. Cowfield was worked down to a fine seed bed so that we did not lose seed under lumps of soil. But I wanted to exploit the fact that after a hay cut, autumn grazing of the regrowth would rapidly multiply the wild flowers. Most seed houses recommended 6 kg/h (4.8 lbs/acre) of wild flower seed to complement the usual 20kg/h (18 lbs/acre) of grass seed. My aim was to test much lower rates:

	kg/ha	gm/sq.m	% of accepted rate
Accepted rate	6.0	0.6	100%
Top half of Cowfield	2.5	0.25	42%
Lower half Cowfield	0.75	0.07	12%

It was not too many years before it was impossible to tell the difference between the two sections of field. Once the crop is established, the grass grows strongly and it gets more difficult, but this is where effective management comes in.

To establish wild flowers on a field scale at 12 per cent of the accepted sowing rate was startling to say the least. It demonstrated that a far lower rate was feasible and enabled us to advise farmers that they could safely reduce wild flower rates to less than half the accepted

Two wild flower seed rates were used here. Half the field was sown at 2.5 kg/ha (2lbs/acre), the other half at 0.75 kg/ha (0.6 lbs/acre). After five years it was hard to tell the difference

rate, thus achieving very considerable cost savings. This undoubtedly helped us get into the wild flower seed business, an extremely competitive market. Soon afterwards we ran a workshop for ten advisors from the Countryside Commission to demonstrate that, with the correct management, the establishment of wild flowers on farmland was feasible. As a result we had growing support not only from the Commission but from all the voluntary conservation agencies as well. It has to be said that chalk soils are one of the easier soils to do this on. Getting wild flowers established on fertile clays and other soils is more difficult and probably requires the grass parasite yellow rattle, but I was only just beginning to discover how effective this parasite was.

THE BUTTERFLIES RETURN

I had only modest expectations that butterflies would increase as a result of the additional nectar. The first of our wild flower seed crops, modest strips of about a fifth of an acre (1000 sq.m) were attracting plenty of customers, but the increase in insects in Cowfield was prodigious. Within a few years Meadow Browns, Marbled Whites and Common Blues had become abundant and we have now recorded thirty species of butterfly. More surprising was the way grasshoppers had spread from the top hedge to the bottom of the field 200m away. Silver Y moths and Burnet moths also appeared in their hundreds.

Wild flowers clearly had the capacity to transform the insect situation, which until then had been universally gloomy. The arrival of unexpected butterflies as well as a great deal of other wild life has not only been exciting but has demonstrated that if you provide the appropriate food plants and nectar plants anything can happen so long as the newcomers can find you. This is an area where we have a huge amount to learn because it is only now with new habitat being created that we can

test the colonising powers of our butterflies and other insects. Before this there was really no means of finding out. Thus when the Small Blue arrived and later the Brown Argus and Dingy Skipper no one even knew where the nearest colonies were.

HELPING OTHERS RESTORE THE COUNTRYSIDE

It was clear that establishing wild flower seed was difficult and that most people needed help. Our first workshops began in 1993 using the farm to demonstrate good practice. They proved very popular. A small visitor centre was opened in 1997 with all the usual nature trails, so that visitors could see the restoration work. We also opened a wild flower demonstration garden showing 150 species, all clearly labelled, and arranged in groups under the common habitats of meadow, hedge, wetland and woodland.

Although the garden is hard work to prepare in the spring it is a wonderful teaching resource for those who do not know a cowslip from cow parsley. I learned quite early on that visitors need to see the wild flowers because if you talk about them in the abstract their faces glaze over in incomprehension or boredom. There is also the problem of introducing people to wild flowers. In the absence of a label, the only other means is one-to-one teaching, which is clearly impractical.

We now have a proven formula for helping anyone who visits the farm and who wants to

These insects were soon numerous where previously there had been none.
top right: Marbled White in dozens; **top left**: Silver Y and **opposite**: Burnet moth in hundreds; and **above**: meadow grasshopper (*Chorthippus parallelus*) in thousands.
An ox-eye daisy makes a perfect sun-bed

The pond, the wood in the background, the hedge and meadow were all formerly part of a 24 h (60 acre) cornfield and we can describe exactly how each one has been created

learn more about wild flowers. There are three levels of interest that we try to cater for.

For the casual visitor. Nature trails showing what we have done on the farm, photos of some of the work and wild flowers, and the wild flower garden with its 150 species.

For the more serious visitor. Half-day workshops, teaching the theory and practice of establishing wild flowers, with a further workshop on propagation.

For individual projects. On site consultancy visits where tailor-made layout, design, and implementation programme with time scales and costs can all be discussed.

The years 1990-96 saw the completion of our tree planting efforts with nearly 5 h (12 acres) planted up (6000 trees) and the hedge restoration or repairs making steady progress with nearly 3 km (2 miles) of hedge attended to. We were now equipped to demonstrate the restoration and management of the key farm habitats of meadow, hedge, pond and woodland and this was hugely important to keep our workshops up to date and for us to be able to show farmers and landowners what to expect if they set out on a major restoration project.

The most recent part of the story was the arrival of Bob Anderson a career farm manager and his wife Rosanne in April 1996. With the increase in wild flower sales a professional approach was needed in managing the growing number of seed crops and the increasing number of countryside features. With their arrival we were able to build up a small flock of breeding ewes and manage our new grasslands more effectively.

New initiatives continue to emerge, usually in response to problems that need to be solved on behalf of our customers, such as the establishment of ancient woodland flora in new woodland plantings. Like so many other aspects of countryside restoration it is astonishing that we still do not have effective methods for re-establishing this important group of wild flowers. We have now made good progress with our own trials.

GOING ORGANIC – BRIEFLY

Another chapter began in 1998, which was to convert the farm to an organic enterprise. Corn crops were successfully grown after two years of red clover / rye grass leys, which built up the fertility, but the wild oats appeared in huge numbers. When corn prices were really at rock bottom, we were able to sell our organic corn at almost treble the price of conventional corn, but there was a cost to all this. Keeping weeds under control on an organic farm requires a huge amount of labour and a simple cropping system. With all our trial field margins, hay meadows, bits of pasture and other

top: The farm wild flower garden with its 150 species has been a wonderful teaching aid and a very useful resource for seed and plant material

above: If creeping thistle did not exist we would probably still be organic

Field margins full of nectar are vital to help insects move around a farm

experiments life was anything but simple. Above all, there is one weed, creeping thistle, which organic methods cannot control.

THE FUTURE

When the Government rolled out its new Environmental Stewardship Scheme in 2005 we reviewed what we were trying to do.

The first priority was to provide wild flower seed for those wanting to restore grasslands in our area. We had in any case kept our seed crops out of the organic scheme. The second priority was to help others restore the countryside through our workshops and advisory work. Well behind these two priorities came growing organic food, so we decided to leave the organic scheme and concentrate on our main priorities. We have joined the new scheme with a plan to create another high quality chalk grassland to rival Cowfield, and make a return to spring corn on the remaining arable land. This will encourage lapwings, and the winter stubbles will provide food for many farmland birds in winter. The sheep on our grasslands will remain chemically free so our organic lamb sales will continue. So it is the start of another exciting chapter, at a time when interest in countryside restoration is growing by the day. It also demonstrates that dreams do some-times come true, even those conjured up in a railway carriage.

SUMMARY

- 1988: establishment of first wild flower field margins and seed crops.
- 1993: Cowfield trial is set up demonstrating wild flower establishment after a corn crop with low seed rates. Butterflies and other insects begin to appear in prodigious numbers. First workshops in wild flower restoration set up.
- 1996: Bob and Rosanne Anderson arrive to manage the farm.
- 1997: a small visitor centre and wild flower collection are opened.
- 1998-2006: Organic crops see us through the period of rock bottom corn prices but the weeds defeat us. Conventional cropping resumes in 2007.

CHAPTER 6

GETTING HOLD OF THE SEED

THE SIMPLEST WAY TO GET WILD FLOWER SEED IS TO WRITE TO THE MAJOR SEED HOUSES OR LOOK UP THEIR SEED MIXES ON THEIR WEBSITES. THEY OFFER GRASS AND WILD FLOWERS APPROPRIATE TO A RANGE OF HABITATS AND SOIL TYPES WITH RECOMMENDED SOWING RATES.

All you have to do is to order. You may wonder where the seed comes from. Is a British source good enough? This is where the plot thickens.

NATIVE SEED SOURCES

Flora Locale is a campaigning organisation, which argues that seed should be as local as possible. Plants and the insects associated with them have evolved over thousands of years in a way that ensures that an insect that depends on a particular plant will find it in flower at the time when it most needs its nectar or maybe its leaves. If the plant comes from a seed further north and flowers two weeks later, it will disrupt the insect's life cycle.

In any case, there is a strong argument that we should encourage local sources of seed so that local genetic characteristics can be preserved.

It is only recently that agencies such as The Highways Agency have decided to go completely British, so there is less concern now that foreign seed will be used on, for example, new road embankments. One particular problem area was French forage crops, particularly birdsfoot trefoil and kidney vetch, which are about a tenth of the price of their native counterparts. Both these species have been developed by the French to stand more upright so that they can be harvested more easily as a crop. Of course, if this seed is incorporated in a seed mix, anyone familiar with British native plants will quickly spot the difference. But interestingly, our native butterflies seem to be quite at home with these foreign forage crops!

It is only recently that wild flower seed producers have existed at all, so unlike trees and shrubs, we can take action before seed from a particular producer is spread all over Britain. Even so, the pattern of seed producers is extremely uneven, with a cluster of producers in Lincolnshire and north Norfolk, several in the centre of England, a few in the West Country and until recently very few elsewhere. There are however problems in that Spanish bluebells are now spreading uncontrollably in some areas with worrying consequences for our native bluebell.

Trees and shrubs have probably been mixed up genetically all over England because hedging plants are now produced by a very few large producers and much tree seed was imported from the Continent in the eighteenth and nineteenth centuries.

above: Spanish bluebells (above) are spreading fast in areas where they have escaped and will out compete their native counterparts (top)

%	Species (seeds/gm)	%	Species (seeds/gm)
1	betony (700)	5	birdsfoot trefoil (500)
7	common vetch (60)	9	cowslip (1000)
7	field scabious (150)	10	lady's bedstraw (1900)
9	lesser knapweed (400)	5	meadow buttercup (400)
1	meadow cranesbill	1	meadow vetchling (100)
5	musk mallow (500)	6	ox-eye daisy (3000)
1	ragged robin (5000)	10	self heal (1300)
8	ribwort plantain (400)	6	common sorrel (2000)
2	yarrow (6000)	7	yellow rattle (300)

opposite: Harebell – the seeds are so small that they would get lost in a combine

pages 62-63: Red campion is a wonderful plant of the woodland edge and seeds prolifically; it is well suited to the garden

above Table of a typical meadow mix of wild flowers

Combining a single species seed crop

Removing grass stems after brush harvesting

Crop grown seed – wild flowers
Most of the large seed houses produce single species crops that are harvested at the optimum time and then all brought together to make up a wild flower seed mix of anything from fifteen to twenty species.

A typical meadow mix of wild flowers is given in the table on page 65 (seeds per gram are in brackets). The percentages tell quite a complicated story:

- Ragged robin has 1% because it has 5000 seeds per gram, and yarrow 2% has 6000 seeds per gram. Many species have less than 1000 seeds per gram, so a higher percentage might be required.

- Betony and meadow cranesbill have 1% for different reasons. Both occur sparingly in meadows (unless, in the case of betony, it is a wet meadow), but meadow cranesbill is expensive to produce, so there is another factor.

- Common vetch comes in at 7% because the seed is large at 60 seeds per gram, so you need more of them.

- Lady's bedstraw has 10% for quite another reason. It is the only common meadow plant that is rhizomatous. It may have evolved like this because it flowers quite late and would therefore find it difficult to spread by seed.

- Self heal has 10% because it is quite a short-lived plant and often behaves more like an annual.

Seed producers have to ensure that this seed remains essentially wild. If a species is crop grown year after year, its wild characteristics begin to diminish. This is why there is a Code of Practice which states that crop-grown wild flowers and grasses can be described as being of native origin to the wild location for up to six generations removed from the wild parent from which the seed was originally collected. One of the characteristics of a plant which plant breeders remove as quickly as possible is slow and uneven germination. They will be looking for quick and even germination, which in the wild could well prejudice a plant's survival if a difficult season came along and all the plants failed. So there is a real conflict here. Although there are dangers in producing crop-grown seed, there is no doubt that it is an effective method of producing large quantities of seed with all the available mechanisation of precision seed drills, combine harvesters and seed cleaners.

It is a sobering thought that when we were collecting seed, often in very small quantities, to set up our wild flower seed crops at the farm, we were scavenging from hedge banks and roadside verges as well as the neighbouring downland. There was very little other local seed that had survived.

Another aspect of crop-grown seed is that wild flowers, which may be common enough in the wild, are not available from seed merchants. This is because the practicalities of growing the crop or harvesting it are simply too difficult and too costly. Several species of the chalk come to mind here. Dwarf thistle is common enough but it is prickly and low growing and there is no way that you could grow it as a seed crop. Small scabious ripens unevenly and the seeds are not viable unless they really are ripe, so this seed is expensive. And then there is harebell characteristic of old downland where the seed size is so small (20,000 seeds per gram) that the whole operation is painstaking and expensive.

Crop-grown seed – grasses
The commercial grass seed business has been growing a limited number of native grasses as well as all the new cultivars that were originally based on the native grasses. Many seed houses use these commercially grown native grasses because they perform as well as the native grasses that you might collect from an old hay meadow. Most seed houses have a simple four- species grass mix for low maintenance amenity sowings and situations such as field margins. Species used for these include:

Common bent
Red fescue
Smooth meadow grass
Crested dogstail
Sheep's fescue

Where a more authentic seed mix is required – for example to create a new hay meadow, a wider mix of grasses is required, such as:

opposite: Our small-scale wild flower seed production is a bit like medieval strip farming

pages 68-69: Gramps Hill near Wantage, Oxfordshire. Brush harvesting on these slopes needed a cool head

If a meadow looks varied rather than uniform, it may have a wealth of grass species even though it has lost most of its wild flowers

Dwarf thistle – impossible to harvest commercially

Restharrow – erratic seed production makes harvesting a nightmare

Common bent	Crested dogstail
Meadow foxtail	Sweet vernal grass
Quaking grass	Red fescue
Meadow barley	Meadow fescue

All these species are available as commercially grown native grasses, but some seed houses are increasingly producing native wild origin grass seed from grasses found in the wild. There is, of course, a cost to this and, at the time of writing, these grasses are more expensive.

Collecting local seed from old meadows or downland

In many parts of Britain there are wonderful surviving ancient grasslands from which seed can be collected. Brush harvesters have been developed in recent years. They use rotating brushes to brush off the ripe seed and collect it in a large box, from where it is laid out to dry and then cleaned. But it is not that simple. Given the long harvesting period of wild flowers, more than one visit will be required in order to harvest seed from a reasonable number of species. There is a key time of year – around the second half of July – when the majority of species can be collected, including nearly all the grasses. Subsequent visits can then collect sufficient seed of the later seeding species, sometimes by hand if the site is not too large.

Stonehenge

This area of 16 h (40 acres) belonging to the National Trust was a cornfield until 2000, when it was sown with grass and wild flowers collected by brush harvester from Ministry of Defence land on Salisbury Plain. It has been harvested regularly to produce seed for converting a number of arable fields to grassland in the immediate vicinity of Stonehenge.

Gramps Hill Site of Special Scientific Interest

Gramps Hill is a superb piece of downland above the village of Letcombe Bassett. The farmer wanted to convert ten acres of arable land at the foot of the down to a wild flower grassland and the obvious place to collect the seed from was the down. The steepness of the down made it only just practicable.

Seed collection took took place on 18th and 19th July 2006 in extremely hot conditions. Because of the heat of the summer, growth was considerably reduced and there was only just enough seed available. A return visit was made on 5th August to collect small scabious, dwarf thistle, and other late species; and again on 25th August to collect lady's bedstraw. The seed was sown on 2nd September 2006.

The current situation regarding local wild flower seed sources can be summarised as follows:

- in central south England, most of the south-east and eastern counties, any land that was not ploughed up to grow corn in the boom years of the 1970s and 1980s is too steep to get a tractor on, which, of course, rules out the use of a brush harvester.
- exceptions to this rule includes the Ministry of Defence land on Salisbury Plain and the Thames Valley meadows around Oxford.
- in the West Midlands, Wales, the West Country and further north in Yorkshire, there are existing hay meadows that can be used for seed collection.

If the seed is being collected from a local high quality site, there is complete certainty that it will possess all the inherited characteristics of the locality. You may well collect local ragwort, creeping thistle and dock seed as well, but this is something that has to be taken into account when the site is reconnoitred in the first place.

The untapped grassland resource

There is however a surprising and largely untapped resource that may have very few wild flowers but the grasses will be varied and original. There are fine hay meadows in many a parish where, over the years, herbicides have removed most of the wild flowers, except the toughest such as meadow buttercup, common sorrel and yarrow. But they may well contain a wonderful array of grasses. In my own parish there are two meadows with ten species of grass, most of which would be very good in a seed mix. These grasses can be brush harvested and then added to local crop-grown wild flower seed to provide a first-class seed mix with a wider range of species than would be possible from brush harvesting on its own.

Using silage or green hay

There are plenty of recent examples where green hay or silage has been spread on ex-arable land in order to establish wild flowers. Most of the seed would have been barely ripe but much of our native meadow seed can ripen after being cut and then establish effectively.

opposite: Greater knapweed is a bold plant easy to identify and collect seed from

Yellow rattle (easy to collect)

Lesser knapweed (easy to collect)

Ragged robin (easy to collect)

Field scabious (difficult to collect)

Small scabious (difficult to collect)

Cat's ear (difficult to collect)

Common vetch (difficult to collect)

Meadow cranesbill (in a class of its own)

Why do meadowsweet seeds form a whorl?

Why did agrimony seeds evolve so that they stick to passing animals?

Why do ox-eye daisy seeds have stripes?

COLLECTING YOUR OWN SEED

This merits a health warning. Collecting seed is addictive, especially wild flower seed! But it is a challenging undertaking and immensely rewarding. There are a number of essential steps that have to be taken:

Compiling a species list

Is there a meadow within reach where you can make up a species list of the wild flowers and grasses? You will need an illustrated Flora but there is no shortage of these today. If a local meadow does not exist, you will have to make do with road verges or other fragments of wild flower areas (remember to get permission from the landowner if these are on private land), but guidance will be needed on a likely list of species, so a visit further afield to a meadow belonging to one of the Wildlife Trusts may be necessary. Most Wildlife Trusts now have websites giving the location of their reserves.

Planning your seed collection

Once you have your species list, you can begin to plan your seed collection. Seed catalogues are again useful because a high price for seed of a particular species will usually indicate that it is difficult to harvest, and a low price that it is relatively straightforward. Start with your own parish and see how many species are available before going further afield. You may find footpaths or quiet roadside verges that you never knew existed, often with steep banks, which on closer inspection turn out to be goldmines of diversity.

You will need to mark carefully where a particular species is when it is in flower and highly visible. It is a different matter when you are looking for a dried-up seed-head. And remember that the roadside flail will suddenly roll up when you least expect it and wreck your collecting plans!

Collecting and collecting dates

Collecting your own seed brings another challenge. You have to familiarise yourself with harvesting dates, of which more later. But there is another problem. Wild flowers have a whole range of strategies for spreading their seeds around. This is known as dispersal. Some species are helpful and more or less wait for you to turn up and tip their seeds out of a neat cup. Others are not so helpful.

Easy-to-collect species: Red campion, ragged robin and lesser knapweed hold their seeds in a cup that opens when ripe, so unless there is a strong wind, the seed will wait for you. Yellow rattle has a pod in which the seeds are packed rather like in an envelope.

More-difficult-to-collect species: Meadow buttercup seeds sit on the end of their stalk and fall to the ground when they turn black. Since they start off green, you have plenty of warning. Common sorrel, on the other hand, is not co-operative: when the disc-like seeds appear, they can all be on the ground if the weather is hot for a couple of days. Common vetch and birdsfoot trefoil are both legumes and have pods, which twist and split when ripe. When this happens all the seeds fall out, so you need an early warning system, which is the change in colour. Common vetch seeds gradually turn black and birdsfoot trefoil dark brown, so in the case of the latter you harvest them when they look like chipolata sausages. Field scabious and small scabious will wait for a while, but after a time the seeds just fall out. Field scabious seed is beloved of linnets, which

Species	harvesting date	species	harvesting date
Common sorrel	18/6	Self heal	5/8
Meadow buttercup	5/7	Field scabious	15/8
Ox-eye daisy	10/7	Lesser knapweed	20/8
Cowslip	15/7	Birdsfoot trefoil	25/8
Yellow Rattle	18/7	Lady's bedstraw	7/9
Common vetch	20/7	Yarrow	11/9

above: Harvesting dates of some of the commoner wild flowers

have a tendency to peck out the key stone (or seed) that holds the seed-head together.

Really difficult species: Meadow cranesbill is in a class of its own. Its seeds are spring-loaded and are catapulted some 2 m (6 ft) away if there is a good breeze. This makes harvesting a nightmare. No surprises that the seed is very expensive. We had two clumps of meadow cranesbill left on the farm. In fourteen years, the seed has travelled some 80 m (90 yds) from the hedge across the field.

The collecting season is a long one but most grasses and many of our meadow wild flowers should be collectable at the end of July (wild flower harvesting dates are given in the table on page 73). Remember once again to ask permission if some of the wild flowers are on private land. My own experience of collecting is that landowners are usually delighted to help if they know that only small quantities are being collected.

Remember also that the number of seeds per gram is a guide as to now much seed you will need of each species. Yarrow has 6000 seeds per gram, while birdsfoot trefoil has 500, so you will need nearly ten times the amount of birdsfoot trefoil seed. Once again, the catalogues of the seed houses give the percentages of the various species so you can calculate the proportion of each species.

Then there is the seed rate. Most seed houses calculate sowing at 3-4 grams per sq.m, with a wild flower element of 20 per cent or 0.6-0.8g/sq.m. This is a sensible rate for small areas.

Drying, cleaning and storage

Collecting can be carried out in anything, such as polybags. Bring it home and dry it in a warm airy atmosphere. You will be surprised how many camp followers appear out of the seed heads. Beware of cooking it in a hot greenhouse. Too much heat will affect germination. Warm temperatures out of direct sunlight are ideal.

Cleaning is a science in itself but much can be achieved with "borrowed" kitchen sieves. The principle of cleaning is simplicity itself. There are two stages:

Stage 1. You need a sieve a little larger than the seed you have collected, so that the seed falls through but the husk and other bits of rubbish remain in the sieve.

Stage 2. You then need a sieve smaller than your seed, so that the seed remains in the sieve and the dust and small pieces fall through.

This DIY operation can get quite sophisticated, with a fan running beside you so that when the seed falls through the sieve during Stage 1, the lighter seed, which may well not be viable, is blown to one side but the heavier seed falls to the ground.

The really important part of the cleaning is to remove the camp followers. It is wise to clean the seed as soon as you can, because the longer you leave it, the greater the danger of some of the insects setting up shop inside your precious seeds. I learned about this the hard way. I had collected some seed of tufted vetch, which to my excitement had reappeared on the farm in a hedge. This is a large seed, with a mere 60 seeds per gram. The pace of life then overtook me and cleaning did not take place for many weeks. The seed was duly cleaned and put in a transparent bag. I was looking at it thinking that the job was well done, when something caught my eye. I peered closer. Two small legs were protruding out of a small hole in the seed. It was some sort of weevil. Alas, most of the seeds had occupants, and all because I had delayed seed cleaning. As a matter of fact, tufted vetch, as well as meadow vetchling tends to pick up travelling companions because they have very convenient seed pods, as well as being some of the larger seeds. Red campion, a much smaller seed, is also prone to insect activity. Cleaning should be done as soon as possible.

Storage should be in a bag (we use self-seal polythene bags) with a mothball added to discourage any extra activity. Being able to see through the bag means that a visual check can be done very quickly. As a general rule, if you do collect seed, it is best to sow as soon as possible.

Multiplying up

This brings us on to the subject of multiplying up your own seed. How long does this take and is it worth it? Some wild flowers such as common vetch are annuals, so they produce seed in the first year. Others, such as meadow cranesbill take a few years before they produce any seed at all. The answer to all this is that if you do not have enough seed, it is far better to sow as much as you can and get the seed in the ground. You can then harvest seed from the first patch and sow additional ground in subsequent years. Additional species can also be added if part of your original list was not available.

SUMMARY

Most agencies in Britain now use native seed.

- crop grown native seen grown by the national seed houses can produce large amounts of seed, but if local seed is required it will have to be collected with a brush harvester or using other methods such as green hay.
- Some species, quite common in the wild on chalk soils, are not available commercially.
- Crop grown grasses are based on commercially grown native grasses but increasingly native wild grasses are available.
- The wild-flower grassland resource in the south of England is now restricted to MOD land on Salisbury Plain, nature reserves and a few small areas in the West Country.
- A surprisingly large untapped resource of hay meadows exists with varied grasses but where the wild flowers have been lost.
- Collecting your own seed is fun and challenging requiring a species list, research into collecting sites, and collecting dates.
- Drying, cleaning and seed storage are all issues of importance.

above: Red campion seed (easy to collect). The process of seed formation is one of the wonders of the natural world

opposite: Red campion

Meadow saxifrage seed is so small that it is best introduced into a meadow as plugs or pots

CHAPTER 7
CREATING A WILD FLOWER MEADOW

WHY DOES SOWING WILD FLOWERS CAUSE SUCH DIFFICULTIES? AFTER ALL, MANY OF OUR GRASS SEEDS ARE AS SMALL IF NOT SMALLER

THE PROBLEM OF SEED SIZE

Wild flowers with a seed size of 10,000 seeds per gram such as wild marjoram can establish well on suitable soils. However, there does come a point when seed size is so small that a seed tray has to be used. This is the case with meadow saxifrage, an old meadow species, which has 50,000 seeds per gram. We have found that if a strong plug is grown in a seed tray and then planted out in the field, colonisation proceeds satisfactorily.

The problem lies in the cost of wild flower seed. It is expensive and because of this it has to be sown at a very low seed rate. The key issue is the number of seeds sown. Amenity grass seed is sown at around 25g/sq.m. Wild flower seed is sown at 0.6g-0.8g/sq.m. This huge difference is compounded by the slow growth of most wild flower seeds and it is for these two reasons that additional precautions need to be taken.

FERTILITY FAVOURS THE WEEDS

Because wild flower seed is slow to germinate as well as being slow to establish, it can easily get overtaken by what we might loosely describe as "weeds," which covers basically everything else that happens to be around. These weeds grow extremely fast because conditions are so different today from, say, a hundred years ago. Farmers have been encouraged to throw huge amounts of fertiliser at their crops. Several chickens are now coming home to roost regarding this practice with unacceptably high levels of nitrates in aquifers and curbs being imposed to avoid further build-up, although it has to be said that the long lag between fertiliser being applied to the land and its appearance in our drinking water means that nitrate levels will rise for many years to come. But this profligate use of fertiliser has affected everything anywhere near the edge of a field, including hedges, ditches, roadside verges and banks, and particularly small river valleys, which now resemble linear nettle beds.

It is partly because our fields are so fertile that it is usually unwise to attempt to add wild flowers to existing grassland. The grasses will simply out-compete any seed that is added. However, there is a way forward using the grassland parasite yellow rattle to control the grasses. It has to flower and set seed, and the autumn re-growth has to be grazed: this is discussed later in the chapter. Meanwhile, the safest method is to sow grasses and wild flowers together: the grasses must not be allowed a head start.

THE KEY WILD FLOWERS

There are a small number of key wild flowers. The list is short but it is essential to get to know these species:

Cowslips. These come first, before the grass starts growing. Cowslip seed has been known to survive years of corn growing and then reappear when corn growing ceased. Seeds sown in the spring will not germinate until the following spring because winter chill is required to break the seed dormancy.

Meadow buttercup, common sorrel and yarrow. The first two follow on from cowslips. All three seem to have survived in many a hay meadow when all other species have disappeared. The leaves of meadow buttercup may be partially resistant to herbicides, and sorrel and yarrow are deep rooted, with the latter having the

above: Seed can be slow to germinate and nothing appears to be happening. A closer look usually tells a different story

pages 76-77: Cowslips will thrive in almost any kind of grass so long as the season's growth is removed and the grass is left short at the end of the year

ability to send out runners so that it can survive from small fragments of rhizomes. All three of these species also have the ability to compete with strong grass growth.

Birdsfoot trefoil, lady's bedstraw, self heal and autumn hawkbit. These are summer flowering meadow species and often survive in lawns, demonstrating their ability to be mown continuously and then flower the moment the mowing stops. Lady's bedstraw used to be so common that pillows and mattresses were stuffed with it in medieval days because of its sweet aroma. It also spreads rapidly by underground rhizomes. Birdsfoot trefoil is important for another reason in that it is the food plant of the Common Blue butterfly (now uncommon). Self heal was a hugely important medicinal plant and autumn hawkbit is a key nectar plant for insects.

Lesser knapweed, ox-eye daisy and field scabious. Ox-eye daisies are colourful and often make a massive display of colour in the early days of a new wild flower meadow. They then diminish to a scattering of plants, which is what you tend to find in the wild. Lesser knapweed is a key nectar plant for butterflies. It often starts in relatively small amounts and can then dominate if it is not cut early enough, hence its name. If it is allowed to flower through until September, it will dominate the whole area. This was why the hay cutting date of 25th July was considered important to keep the various species in balance. Field scabious has high quality nectar for butterflies but the seeds are popular with finches and only seem to germinate well on thin chalky soils.

Ribwort plantain and yellow rattle. Ribwort plantain was also called ribgrass because yellow rattle used to wipe out all the grasses leaving ribwort plantain as the only green looking, grass-like plant for the cattle to eat. For this reason it is an important element in the wild flower meadow. Yellow rattle is the key to grass control and spreads rapidly so long as a hay cut is taken with autumn grazing.

Plants of the chalk. There is a long list of plants which can be found on the chalk but rarely on other soils. Kidney vetch is a great coloniser of bare chalk and establishes quickly. Salad burnet can grow in great quantities and almost act as ground cover. Common St John's wort has tiny seeds but is well able to colonise bare chalk, as is harebell with even smaller seeds. Common

Birdsfoot trefoil

Common sorrel

Autumn hawkbit

Cowslip

Field scabious

The key wildflowers

Lady's bedstraw

Ribwort plantain

Self heal

Lesser knapweed

Meadow buttercup

Ox-eye daisy

Common vetch

Yellow rattle

Yarrow

The golden glow of meadow buttercups on an early summer's evening used to be one of the sights of the countryside. This meadow was in the corner of a huge cornfield some fifteen years ago

rockrose is slow to get going and then behaves more like a small shrub. Other species include greater knapweed and small scabious; and clustered bellflower, wild thyme, and wild marjoram shown on page 84.

STARTING AFRESH OR MAKING REPAIRS – THE BIG DECISION

There is no escaping the big decision of whether to start afresh with a clean seedbed or whether to carry out repairs, perhaps over a few years. At one extreme is the field of stubble or ryegrass and clover, where 100 per cent new seed is clearly going to be required. At the other extreme is an area of old meadow grasses, with many varieties and with possibly a few wild flowers as well, where there is too much diversity of local interest to lose if you start afresh. The problem is that between these two extremes there are many grey areas that can be summarised as follows:

Start afresh: with 100 per cent hybrid clovers/perennial ryegrass and stubbles.

Repairs: with 100 per cent old meadow grasses, with or without a few wild flowers. Species that tend to survive include meadow buttercup, common sorrel and yarrow.

Grey areas: A key question is the amount of perennial ryegrass. As a general rule, if the ryegrass constitutes more than 30 per cent of the sward, it is better to start afresh. However, remember that perennial ryegrass occurs naturally in most old hay meadows comprising 5–10 per cent of the grasses. If you give an old meadow generous helpings of nitrogen fertiliser, the ryegrass expands and soon dominates, out-competing all other species. The wild flowers are the first to go, followed by the less vigorous grasses. In theory, you should be able to reverse the process but the timescale of thirty years plus is daunting. We cannot wait that long.

If there are a few surviving wild flowers, such as meadow buttercup, sorrel and yarrow with, say, 30 per cent perennial ryegrass and a reasonable variety of other grass species, then it may be worth going down the repair route. If there are also cowslips, lesser knapweed, lady's bedstraw, ox-eye daisy or birdsfoot trefoil, you should definitely choose the repair route. Some grasses are more indicative of quality than others.

There is a further point: we do not really know the effect of tiny differences in the time of flowering of local grasses on the animals that have always lived in that particular locality. There is a huge need to retain local species if we possibly can. But it comes down to making a decision to start afresh or make repairs. Starting afresh is much the most complicated so I will discuss this first.

This is also the point where a division has to be made between gardeners and farmers. Of course it is a gross generalisation but farmers tend to have the appropriate equipment and/or animals to graze. Gardeners often have awkward areas where equipment does not have access and they may become paranoid at the thought of a flock of sheep anywhere near their roses.

opposite: The paler-flowered form of sainfoin was introduced as a fodder plant in the seventeenth century. Now widely naturalised, it occurs on Salisbury Plain. It has a large, bold seed, is easy to establish, and is more upright than the almost prostrate native form

STARTING AFRESH IN THE GARDEN

Areas too small for a tractor (and today's tractors are huge unless you have an ancient grey "Fergie" or an old red Massey Ferguson) often need much ingenuity to complete the essential operations. Most garden areas are already in grass, and the area may have been under managed, so there will be dormant seed that needs to be taken into account. The first decision is when to sow, in spring or autumn.

Wild flower seed ripens naturally in the late summer and will have fallen to the ground by the autumn, so this is the time of year that it is conditioned to germinate. There is, however, a

Clustered bellflower, a beautiful blue flower

Wild marjoram, arguably the best nectar source on chalk

Wild thyme famous for its carpet effect

more important point. When wild flower seed falls to the ground, it remains on the surface, where it is perfectly able to get on with its life, with warm ground and falling temperatures. Sowing in the autumn with a fine and well-consolidated seedbed and leaving the seed on the ground and then rolling it, means that there is nothing much else that we can get wrong. But if we sow in the spring, we cannot adopt this practice, because seeds cannot be left on the soil surface at this time of year. Temperatures are rising and the seed has to be lightly covered to protect it from the fierce heat as it grows. We have all experienced mini heat waves in April or May. So this means that the seed has to be covered, which can cause difficulties.

There is simply a lot more that can go wrong if you sow in the spring. Of course there are always exceptions, such as low-lying land that floods in winter and has to be sown in the spring.

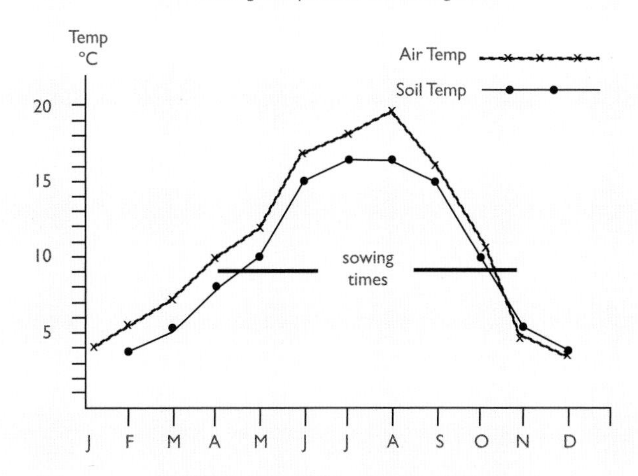

above: Diagram of temperature and time of year re autumn and spring sowing: Autumn is when most wild flower seeds germinate naturally and when they can be left on the soil surface. Spring sowing is more complicated

Red fescue

Crested dogstail

Common bent

Meadow barley

Sweet vernal grass

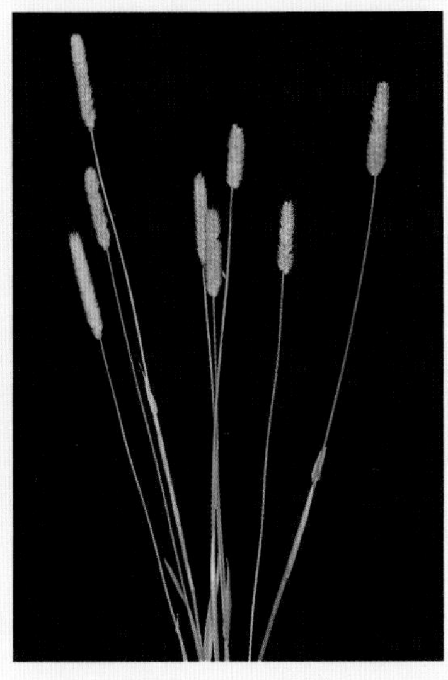

Small leaved timothy

Quaking grass

Yellow oat grass

Grasses for the wildflower meadow

Grasses to avoid in a wildflower meadow

Yorkshire fog

Perennial rye grass

Cocksfoot

False oat grass

Essential Operations:

May	Spray the old grass with a herbicide
June	Cultivate with a rotovator, roll to consolidate
July/August	Harrow or rake the weed flush
	Repeat several times
September	Give the final spray
	Sow and roll

Spray the old grass

If you dislike all chemicals, you have my sympathy but there is a price to pay in time. You may need a lot of it. Grass destruction can only be carried out in the summer months by repeated rotovating or cultivating and the sun has to shine for reasonable periods. If the weather is against you, you will achieve nothing. Alternatively, you need to get hold of a 15-litre knapsack (which can be hired) and some herbicide, and spray the grass and everything else on a fine still day when there is active growth and the grass is 12 cm (4-6 in) high. "Spray in May and go away" is a good adage. You should allow two weeks for the chemical to take effect.

Cultivations

Given the small size of seeds and their very low numbers, it follows that the condition of the soil where they are deposited is of considerable importance. It is simply no good creating a cloddy seedbed and expecting the wild flower seed to establish itself. Seeds will get "lost" under clods the size of golf balls or larger and never see the light of day. A really fine seedbed is the objective with, if possible, clods no larger than marbles.

Rotovators can be hired, but here again there are pitfalls for the unwary. Do not use a rotovator on heavy soils such as clays in wet weather otherwise you will smear the soil and damage the soil structure. After rotovating, get out the rake and remove all the dead material, which can be put on the compost heap so long as there are no couch rhizomes: these should be burned. This can be very hard work, but the quality of the seedbed depends on it.

The implication of all this is that seedbeds are weather dependent and cannot be summoned up at short notice. So much in today's culture can be produced almost instantly that it can come as a shock to be told that you need to allow a growing season to produce your seedbed, but that is the reality. If the job is rushed, with a decision taken in late August to convert a rough piece of grassland into wild flowers with sowing in September or October, it can easily end in disaster. Herbicides will be ineffective if the weather is hot (grass needs to be growing actively to absorb the chemical) and the ground will be too hard for the rotovator to work. The job needs to be planned so that the work can take place under favourable conditions.

Rolling

You then need to roll the ground to consolidate it and you are ready for the weeds to germinate. This germination is known as a weed flush, but the weeds will only grow if the soil does not dry out and remains as moist as possible. Rotovating leaves the soil very open and hollow (when you walk on it your feet sink in) and poses a real risk that the ground will dry out and nothing will happen at all. This is where rolling is so important, because if the ground is consolidated, the damper, lower levels of the soil will feed moisture to the upper levels through capillary action. If you do not have a roller, drive the ride-on mower up and down. The wheels are a reasonable substitute.

Harrowing or raking the weed flush

This is all about dormant seed, which is divided into perennials such as perennial rye grass and Yorkshire fog and annuals such as groundsel, speedwells, poppies, fat hen and many others, which will be plentiful in old kitchen gardens. Perennials need to be killed off, whilst the annuals can be mown or strimmed through the first season and will not reappear because they need cultivated ground to germinate.

It is when you convert one kind of grass to another kind of grass with wild flowers that the problems of dormant perennial seed occur. Most areas of old grass in orchards or derelict sites are dominated by false oat grass or cocksfoot, and it is absolutely essential that these be reduced.

If you are converting an area of coarse grass to fine-leaved grasses, there is this additional step that is not immediately obvious. Step 1 is to spray off the existing vegetation, which is clear for all to see; but Step 2 is to reduce the dormant seed, which you cannot see.

The way in which a piece of ground has been used in the past leaves a lasting legacy of weed seeds so it is fairly easy to predict where the problem areas are going to occur. Kitchen gardens, old well-managed lawns and orchards are low risk but derelict areas with rank grass are high risk.

The aim is to reduce the dormant seed in the top half to one inch of soil as much as possible. If the seed is not going to germinate, it can remain where it is, deep in the soil, but if it is within reach of light, it will grow and interfere with the precious seed you are about to sow. Reducing the dormant seed can only be achieved by encouraging it to germinate. The first green haze is a good start. What you need to do now is to wipe out the seedlings and bring up any more that are lurking in the vital inch of topsoil. On a small scale, this is achieved by a light raking (preferably with a wire rake) on a dry day to uproot all the seedlings so that they will be killed off by the sun. On a larger scale a harrow frame pulled behind a ride-on tractor/mower will achieve the same result. Some extra weight on the harrow frame may be helpful. It is essential to keep the top inch moving and not to go too deep.

After the first operation against the weed flush, you are ready for a repeat. For this you need a little rain, which may mean waiting several weeks, but this does not matter as long as you have given yourself plenty of time in the first place. How many times should you go for a weed flush? Two or three times if you have enough energy. But when sowing time arrives in the first week of September a different tactic is required.

Giving the final spray

Repeated attacks against the weeds reduce them, but there are always more. When September draws near, if the ground really looks clean, you can sow without further ado. But if there appear to be more weeds germinating, you need to apply more herbicide (a half rate dose will be sufficient) and you can then sow the next day. You can in fact sow within the hour, if the spray has dried. This technique of spraying before sowing is really helpful if you have only been able to achieve a single weed flush. But this is where there is a golden rule.

When you have sown the seed, do not rake or disturb the soil apart from rolling, otherwise more dormant seed will be brought to the surface.

Sowing and rolling

You will have noticed that I suggested autumn. Wild flower seeds are best left on the soil surface at this time of year for the simple reason that if you attempt to cover them, you may end up burying them too deeply so that they decline to germinate. If you decide to sow in the spring, it is essential to cover the seed with careful raking. If you are using a ride-on mower to harrow a larger area, the ground must be consolidated otherwise the harrows will pull too much earth over the seed and it may not reappear.

There is also the problem of creating a seedbed in spring after the winter rains have compacted the ground. This means some sort of cultivation, but there will be no means of reducing the dormant seed because the soil temperatures will be too low for anything to germinate. You will have to sow blind as it were, not knowing the severity of the dormant weed situation.

Wild flowers and grass need to be sown together. Why do you not just sow wild flowers alone, since so many problems seem to come with the grasses? The problem about sowing wild flowers on their own is that for most of the winter, you would be looking at bare earth since many wild flowers disappear completely in the winter. If there are people or animals running about on an area without grass, it soon turns into a mud bath. But there is also a more important issue. Nature abhors a vacuum, and where there is bare earth, seeds of unwanted grasses and other weed species soon arrive and cause problems.

To return to sowing or, to be strictly correct, "broadcasting," because that is the process that leaves the seed on the soil surface, how do you go about it? If you have an area less than the size of a tennis court, there are two problems that will confront you:

If cocksfoot and false oat grass dominate a site, that tells you all you need to know about its past management

- wild flower/grass seed rates at 3-4g/sq.m (20% wild flowers, 80% grasses) are extremely low, and you will run out of seed!
- How do you know where you have sown the seeds?

Sowing seed for the first time is a great adventure, but there can be problems: running out of seed is one of them – and a somewhat embarrassing one. It smacks of incompetence. If it is any consolation to the reader, I used to run out of seed constantly! Then I discovered sawdust, or it could perfectly well be sand. Mix the precious seed with five or ten times the volume of this additional material. You are at once back in control: you have plenty of material to sow and you can see where you have been, because both sand and sawdust show up nicely on the soil surface.

As an added refinement, you divide the area to be sown into four quarters and the heap of material to be sown into four buckets, and you then broadcast the seed onto one of the quarters. If you make a complete hash of it, all is not lost. An amazing number of customers telephone us in the most abject terms, apologising for running out of seed and asking us to send them more!

Once the seed is on the ground, it needs to be rolled, but if it just keeps raining, it does not matter because the soil surface is not going to dry out. Rolling is important when the soil surface dries and moisture needs to be brought up from lower levels.

Over-seeding

If you are sowing several areas and you want variation in colour or even variations within a single area, over-seeding is the method to use.

After sowing a base of wild flower seed mix, some of the best plants to achieve splashes of colour are:

- *Meadow cranesbill (blue)*: it is slow to establish, but once it is away it is spectacular. The seeds are large and tough and the clumps will expand.
- *Tufted vetch (blue)*: this has a similar seed size to meadow cranesbill but the seeds seldom germinate in the wild. This may be because the seeds start off in pods which get parasitized by a grub.
- *Field scabious (powder blue)*: this will only spread from its seeds if it is on thin soils.
- *Wet meadow or late species for damp areas*, which can be sown together or separately, such as *betony* (purple), *devilsbit scabious* (deep blue), *fleabane* (yellow), *meadowsweet* (white), *great burnet* (deep purple). Fleabane is the easiest to get going as it spreads from underground rhizomes. Great burnet requires damp ground.
- *Cowslip (yellow)*: although there are cowslips in most wild flower seed mixes, there is an argument for having a spectacular display of them where they are easily seen in order to brighten the early part of the year. This area can then be kept short later in the year with mowing resuming some time in June.

above: Common poppy; **centre**: Common field speedwell
Strimming or mowing through the first season will prevent these plants from dominating new seed

Rotovated pathway

Weed species in the derelict meadow

If docks are widespread, serious questions need to be asked as to whether this is the best site

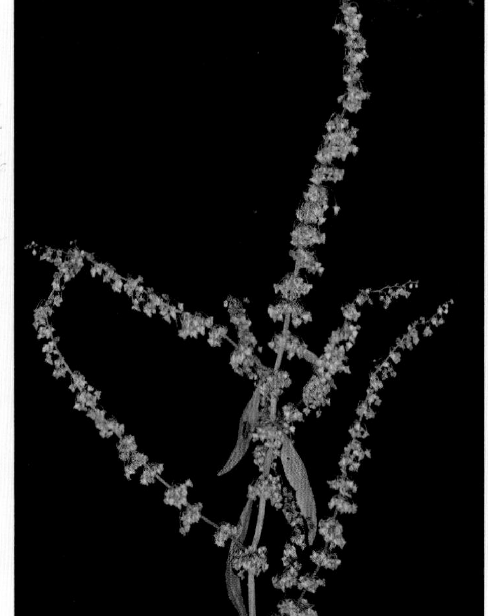

Dock seeds last for ever in the soil

Fat hen is abundant in kitchen gardens

Chemicals

There is a view that the use of chemicals will somehow prevent wild flowers establishing or that it will make the whole process more difficult. I have found no evidence of this. Removing difficult weeds or unwanted grasses is a one-off event and should be done as effectively as possible. I would add here an example of a derelict field of 0.2 hectares (half an acre) that had so much bramble encroaching across it that the meadow was almost lost. The scrub was cleared in wet weather and the ground severely churned up. When it was all levelled, we waited to see what would happen. To our dismay, it grew docks, thousands of them, wall to wall! So we sprayed the docks with a fairly strong selective herbicide. The ground was prepared in the autumn and left over winter. We knew that the winter chill would break the dormancy of all the dock seeds still lying near the soil surface. We kept a close watch on the field and when the docks started to chit, we walked the field with knapsacks and sprayed it again. We managed to get most of the docks. Two years later, over 200 common spotted orchids appeared at one end of the field. I think this tells us that wild flowers are quite tough and that orchids are really tough!

DERELICT SITES

It is almost inevitable that the site chosen for this brave new experiment of sowing wild flowers will be a semi-derelict area at the end of the garden, where nothing has happened for years. I shall assume the worst: the grass is rank; there is a patch of docks, a large clump of creeping thistle, and a huge bed of nettles with roots like hawsers, and plenty of hogweed. This sort of area will take at least two seasons to clean up. Let me explain why. Once again, there is the problem of dormant seed, this time nettles, together with a very small window to deal with the rhizomes of creeping thistle.

First winter – clear the decks, dig out brambles, and cut everything else down to the ground with a strimmer or brush cutter so that in the following season growth can proceed unimpeded by dead material.

First growing season – in May: attack the docks and nettles with selective herbicides and the grass with a general herbicide. Note that creeping thistle can only be dealt with in May/June when it is about 20-25 cm (8-10 inches) high. In September: a follow-up attack on docks and nettles. In October: rotovate.

Second growing season – rotovate or harrow whenever the area greens up. Sow in September.

PARTICULARLY DIFFICULT SITES

If 50 per cent of the area is infested with docks, find an alternative site.

If all the docks are in one part of the meadow, avoid sowing any seed there and give yourself another season to clean it up. Alternatively,

Creeping thistle – to eliminate it, it is essential to use a chemical that will get down into the rhizomes

Nettles – dormant seed will go on appearing for several years

Hogweed – it looks scary, but seeds last barely a year

sow grass only at the end of first growing season and use a selective herbicide to remove the docks and thistles that appear in the second season. Plant plugs or pots later in the season.

MEADOW REPAIRS IN THE GARDEN

The first action is to mow less to see if there are any wild flowers around. I visited a customer some years ago who wanted to introduce wild flowers to a grassy bank that had always been mown. It looked as if it might be full of wild flowers that had never had the chance to flower, so I suggested a pause in the mowing so as to see what might appear. Some weeks later I received an ecstatic telephone call saying that the wild flowers were wonderful and the grass was also full of orchids!

There are two other points to bear in mind: if there is a part of the area with thinner soils, always choose this in preference to areas with deeper soils. If you choose an area of lawn that has always been well managed, dormant seed should not be a problem because only the finer-leaved grasses survive regular mowing. The key to adding wild flowers into fertile soils is to establish yellow rattle, which weakens the grass growth and allows the wild flowers to compete. Whether you add seed or plant plugs, yellow rattle can be very helpful. If it does not persist because of the difficulties of organising grazing, early mowing through to the end of May will help take the steam out of the grass growth and allow the wild flowers to compete.

Seed

Sow in the autumn. Mark out the area to be repaired, mow as close as possible and then rake hard to expose 30 per cent bare earth by early September. Light rotovating will achieve the same by mechanical means. Sow wild flower seed without the grass at 0.5-1gram per sq.m, and yellow rattle at 1 gram per sq.m, bulking up the seed, as mentioned above, with sand or sawdust so that it is distributed evenly and you can see where you have sown.

Roll or tread in well.

Keep the area rough mown down to 15 cm (6 ins) through the following year. This avoids cutting too much off the top of yellow rattle, which is an annual.

A variation on the above is to spray off 1 m (3 ft) pathways and thoroughly rotovate, then spread the yellow rattle seed, with additional wild flowers if required. This is a useful method for neglected old meadows with patches where the wild flowers have been lost.

Plugs

Plant in spring or early summer. A plug is a small wild flower seedling grown in a cell or plug tray, which is not only very much cheaper than a wild flower grown in a 9 cm (3.5 in)pot but also much easier to handle. Common sizes of plugs are 40 ml. The downside of plugs is that given the much smaller size of plant, their establishment is more difficult.

Rules for plugs
Plant in the spring when growth doth begin.
Do not plant in the autumn when growth ends.
40ml plugs are small enough.

Plugs – different growth rates require different planting densities (common sorrel, meadow buttercup and yarrow will probably not be required)

6 per sq.m	**8-9 per sq.m**
Knapweed	Lady's bedstraw
Field scabious	Birdsfoot trefoil
Meadow cranesbill	Cowslip
Autumn hawkbit	Ragged robin
Self heal	Ox-eye daisy

Common vetch is an annual and can only be introduced as seed.

After a long hot summer, it is tempting to plant plugs when the ground at last gets moist. But if you think about it, growth ceases in early October, and the small plug then has to survive at least five months when there is no growth, the ground is cold and wet, and no end of hazards abound including bored rabbits looking for a therapeutic occupation.

So plant in the spring, when the ground is warming up. If rabbits are a threat, 1 m (39 in) lengths of rabbit wire can be pegged down over the plugs. We used this method when planting out 9cm (3.5 in) pots of horseshoe vetch to stop the rabbits digging them up.

The preparation and management of plugs is important:

- Give each plug clean ground around it, using a herbicide if necessary beforehand, and plant plugs in groups of 10-12 with 3 or 4 wildflower species in each group. Add yellow rattle seed at a rate of 1gm per sq.m to control future grass growth.
- Plant with a trowel or dibber and press in firmly.
- Keep plugs weeded during the first season, and ensure that surrounding grass does not shade them out. Water well before planting and only water again if there is a drought.
- Allow plugs and surrounding grass to seed into bare earth at the end of first season.
- Cut back to 10cm as for rough grass at end of first season.

Vigilance is required at every stage because you never know what lurks beneath the soil surface. We introduced plugs into an area of grassland by using a herbicide to clear 1 m (39 in) circles into which the plugs were planted. When we next returned to the site, the circles had been completely taken oven by creeping buttercup, which had been lying around as dormant seed.

Always bear in mind that buying plugs is a cost effective way of getting hold of a plant. If you decide that you need a larger plant, it is a simple matter to keep the plug for 6 to 8 weeks and grow it on.

SIMPLIFYING GRASSLAND MANAGEMENT

If you find yourself with an orchard or a new area of grass bought from a neighbouring farmer, which grows uncontrollably, remember that it may well have started off its life producing three cuts of silage per season for the farmer. The hapless owner of this type of grassland will spend hours on a ride-on mower accumulating a mountain of mown material and exhausting him- or herself in the process. Here again a decision needs to be taken. Do you really want this type of grass? Why not replace it with slow growing fescue grasses? If you were going to do this, it would be a good opportunity to add a little wild flower seed as well, or at least some cowslips.

THE FARMLAND CONTEXT

Arable land

The issues confronting farmers are very different. Most areas being converted to grass

If new grass is being established, the opportunity should always be taken to establish wild flowers also

and wild flowers are ex-arable, where any remaining perennials will have been sprayed out of existence by years of agro-chemicals. Thus, the issue of dormant seed is a minor one and simply concerns the mowing or topping of annuals in the first year. This greatly reduces timescales in that when the combine has left the field, the stubbles (assuming the straw has been removed) can be lightly cultivated or disced, encouraging annuals to germinate. When the field greens up, it can be cultivated again or sprayed with a herbicide. Ploughing may be necessary if there are a lot of residues from the previous crop or if the weed burden is considered serious, but this will involve more work to consolidate the seedbed.

Sowing or broadcasting is usually carried out with a standard drill, but with the drill adjusted so that the seed is simply dropped onto the soil surface. Some farmers use fertiliser spreaders or spinners. Cambridge rolls complete the job. This is all normal practice. Some farmers will be tempted to "direct drill" the seed into the stubble. This is a technique where all cultivations are avoided, and a heavier drill cuts a slot into the surface of the stubble field. The seed is dropped in the slot and small tines provide a light covering of soil over the seed. This tends to bury the seed too deeply and the results are usually disappointing.

Seed rates are somewhat more flexible when sowing larger areas if only because of the cost implications. Farmers often take a longer view, especially since they will probably have animals to graze the new grassland. The grazing animal, as explained in an earlier chapter, has the unique ability to not only graze the sward, but to tread in the newly shed seed, something the ride-on mower finds rather difficult! Treading in newly shed seed permits sowing at lower seed rates. Simply allow a few years for the grazing animal to help multiply up the seed in the field.

Reduced seed rates for farmers

Grasses	20 kg/ha (18 lb/acre)
Wild flowers	2.5 kg/ha (2 lb/acre)

This can be reduced to 1.25 kg/ha (1 lb/acre) on chalk, or less fertile land where fine seedbeds can be achieved.

Cholderton Estate, Hampshire

Probably the best example anywhere of a cornfield converted to wild flowers is in West Hampshire, where Henry Edmonds used to grow good crops of spring barley. The 15-acre field was sown to a grass seed mix twenty years ago and Henry has been adding in species ever since (unlike Magdalen Hill Down mentioned in chapter 3 where all the species were sown at the outset) so it is an excellent example of continuous enrichment. Yellow rattle has played an important part in keeping the grass competition under control. What is particularly striking is the number of so-called difficult species which are not commercially available. These have been introduced as seed in very small quantities and have spread widely through appropriate grazing and include autumn gentian, eyebright, wild thyme, restharrow and dwarf thistle. In addition, horseshoe vetch (food plant of the Chalkhill blue butterfly) and devilsbit scabious (food plant of the Marsh fritillary butterfly) have been planted as 9cm pots. The Chalkhill blues have arrived and the Marsh fritillaries have yet to do so. The field is a stunning sight in July and August.

As stated earlier, wild flowers should always be sown at the same time as the grasses. If the grasses are sown first, it is too easy for them to get ahead and out-compete the wild flowers. However, there is one situation where the grasses can usefully be sown in advance. This is where there are steep, chalky slopes at the foot of downland from which corn growing has retreated. With autumn sowing there is a risk that the autumn rains could wash the wild flower seed down the slope. If the grass is sown the previous season at half to two-thirds the usual rate, a matrix will be established that will help the wild flower seed to stay in place. Because of the chalky soils, the grass is unlikely to out-compete the wild flowers.

Grassland

Converting a field of perennial ryegrass and clover to wild flowers is a much sterner test

A massive amount of white clover emerged from the seed bank after the field had been rotovated in order to add wild flower seed

White clover

Autumn gentian now abundant in Henry Edmond's field

Twenty years ago Henry Edmond's field was producing barley. Today it is one of the best examples of a diverse chalk grassland

top: At Hurst water meadows a power harrow was used after a close cut with the flail to prepare a seed bed for additional wild flower seed. Even after this treatment large amounts of dormant grass seed appeared

above left: This ancient set of discs did a superb job preparing the seed bed for additional wild flower seed at Dundridge meadow

above: Dundridge meadow after the new seed had been added

where the required timescale begins in May, as set out at the beginning of the chapter. If you spray, plough, create a seedbed and sow your new seed without taking steps to reduce the dormant seed, you run the gravest risk of ending up with the same field of rye grass that you started with. A full growing season is needed to reduce the dormant seed. Ploughing will be necessary to bury all the dead grass. Seed rates should be as above.

The jury is still out on other methods of introducing wild flowers into existing grassland. Compacted soils often require heavy implements such as discs to break them up sufficiently for seed to gain access to the soil surface, and there is the major problem of competition from the existing grasses. This is an area where trials are badly needed. Our experience here is very limited, but conclusions can be drawn.

At the farm we lightly rotovated part of a field on a chalk bank after grazing the grass quite hard. Wild flower seed was broadcast at the rate of 2.5 kg/ha (2 lbs/acre) in September 2006. Establishment has been successful with all the sown species being observed in Summer 2007, although white clover has been activated from the seed bank and we will have to manage this with care. Clover responds vigorously to cutting or grazing so it will be a balancing act not to graze too hard (stimulating the clover) whilst grazing enough to prevent the grass from out competing the new wild flowers.

Hurst water meadows, Oxfordshire

7 ha (18 acre) Hurst water meadows are located near Dorchester-on-Thames. These fertile meadows had experienced years of neglect resulting in dominance of cocksfoot, Yorkshire fog and false oat grass. In September 2006, 3 ha (7.4 acres) were cut very close with flail cutters and all the cut material collected and removed, after which they were power harrowed and then sown with 75 per cent yellow rattle and 25 per cent other wild flower seed.

Germination of the new seed was delayed by a completely dry month in April, and it remains to be seen just how successful wild flower establishment will be. What was very significant was the level of dormant grass seed that returned after the power harrowing had been completed.

Dundridge meadow, Hampshire

Pete Potts, a senior ranger for Hampshire County Council Countryside Service used discs to prepare the ground before sowing wild flowers along one side and in the centre of a small field which was on loam over chalk at a similar rate to the above with subsequent grazing. The plan was for the wild flowers to then spread across the field. Establishment has been excellent as has been the subsequent spread. Discing was a better method for opening up the sward since rotovating left the surface rather rough.

A list of the species successfully established is as follows:

birdsfoot trefoil	common St John's wort
cowslip	greater knapweed
hoary plantain	kidney vetch
lesser knapweed	meadow buttercup
lady's bedstraw	ox-eye daisy
salad burnet	self heal
wild carrot	wild marjoram
yarrow	glaucous sedge
tufted vetch	devilsbit scabious
yellow rattle	red clover
small scabious	meadow cranesbill
field scabious	wild basil

Dundridge meadow three years later

If the grasses are suitable, it would appear perfectly feasible to introduce wild flowers into grassland on thin chalk soils, but it becomes increasingly difficult as the soils get more fertile and the grasses get stronger. Grazing is a key part of the process and we have much to learn as to how hard grazing needs to be without damaging the wild flower seedlings. It seems very likely that the only way to add wild flower seed into fertile meadows may be to cultivate strips, spray off the grass germination and then sow the new seed, allowing it to spread into the rest of the meadow over time.

You may have a grass field devoid of wild flowers that you are considering diversifying. If the field has not been actively managed, wild flower seed may exist in the topsoil. It is always worthwhile harrowing the field really hard, then grazing it hard and harrowing again in the autumn to see if anything comes from the seedbank during the following growing season.

If the grasses are reasonably benign there is a straightforward procedure for introducing wild flowers together with yellow rattle into an existing sward:

Graze really hard in late August.

Harrow hard or disc to bring up 30 per cent bare earth.

Broadcast yellow rattle seed at a rate of 1.25 kg/h (1 lb/acre), with additional wild flower species at 2.5 kg/h (2 lbs/acre).

Put sheep or other grazing animals back in the field to tread in the seed. If there is some moisture in the ground, the treading will be more effective and the remaining grass will get more effectively trampled.

Take a hay cut the following year in the usual way. A hay cut must be taken at least every other year to allow the yellow rattle to get established

SUMMARY

- Wild flower seeds vary in size from 60 grams to 50,000 grams
- A short list of about 20 key wild flowers is essential knowledge
- Starting afresh or repairing your meadow needs some knowledge of grasses
- Dormant seed lies around in the soil in huge quantities and must be reduced
- Sowing and over-seeding
- Derelict sites
- Meadow repairs: seed and plugs
- Farmland: converting arable land and grassland

CHAPTER 8
ESTABLISHING AND MANAGING THE MEADOW

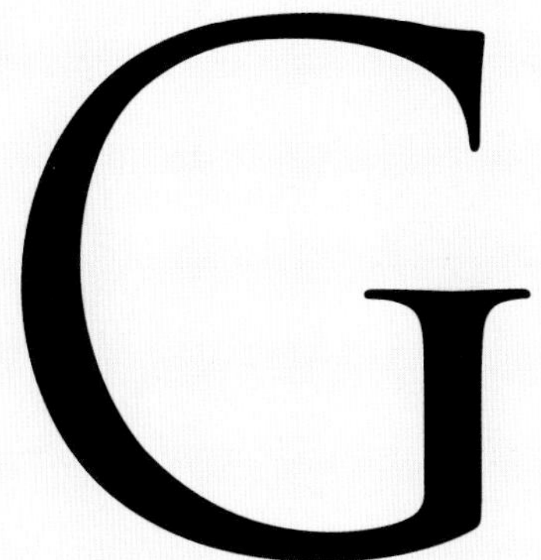

GETTING THE MANAGEMENT OF THE MEADOW RIGHT IS MORE DIFFICULT THAN SOWING IT. HOW OFTEN HAVE YOU HEARD THE COMMENT, "IT ESTABLISHED WELL BUT HAS GONE DOWNHILL EVER SINCE."

We constantly need reminding that our most diverse habitats are the result of intensive management, and that includes the hay meadow. There is also the added point that most of our soils are much more fertile than the conditions to which wild flowers originally adapted and this makes their management even more challenging. Additional fertiliser is definitely not required.

The management process can be divided into three phases:
Year I: Establishment
Years 2-6: High initial fertility
Year 7 plus: Lower fertility and the longer term

YEAR I – ESTABLISHMENT

There is a major psychological problem in the first year. Most seed houses advise that everything should be kept mown down to 4-6 inches all through the first season. Surely they do not mean it. Well, they do! It is all about competition, which in the natural world is ferocious.

Competition

In 90 per cent of soils there is dormant annual seed as discussed in the previous chapter; it is particularly abundant in old kitchen gardens, ex-arable land, and on derelict sites. If the seasonal growth is not kept reasonably short, the annuals will outgrow the wild flowers that are struggling to get established, especially those that develop quite slowly. Two species illustrate this well: both cowslips and lady's bedstraw grow very slowly. At the end of the first season, they are less that 5cm (2 in) high. By this time a plant of fat hen, for example, is 1.2m (4 ft) high and not only removing light, but sucking out all the moisture from around the slow growing perennials, as well as preventing moisture from reaching them from above. Poppies are also a threat because they are big and bushy and there are so many of them in many soils.

The solution is to mow or strim two or three times during the first season. For the garden meadow, the new area should be treated like

pages 98-99: This scene is decorative but dangerous. A neglected game crop has been colonised by true weeds, sow thistles (yellow) and scentless mayweed (white). These are annuals or short-lived perennials and will pave the way for the worst weed of all, creeping thistle

Topping needs to be high enough to avoid removing too much of the yellow rattle

Yellow rattle with wheat in John Wheeler's field. The rapid growth of annuals would overwhelm the wild flowers without regular topping or mowing

Yellow rattle seedlings established well in the crop residues that protected the soil from the extreme dryness in spring 2007

opposite: Because this field margin has not been cut often enough (minimum once a year) lesser knapweed is beginning to dominate and diversity is decreasing

opposite: Black grass is a very competitive annual weed of arable crops that will not re-occur once the perennial wild flower grassland is established. However, it would be best to reduce the amount of dormant black grass seed before sowing the new

rough grass. If you have several acres, then use the tractor and topper on a dry day so that the cut material blows away. Do not on any account do what the roadside flail does, which is to wait for a wet day and cut when the vegetation is about 1 m (3-4 ft) high. This leaves a heavy mulch, and effectively suppresses all growth of small plants. It can come as a surprise that the growth of annual weeds is a one-off event, but annuals need the ground to be disturbed or cultivated for germination and growth to be triggered; if the ground is not cultivated, the annuals do not reappear. This is why it is not possible to have poppies in your meadow. Because they are annuals, you would have to cultivate the meadow every autumn!

Competition is usually the reason why a plant is scarce in some places yet common in others. Harebells are a wonderful example. They are common in parts of Scotland but scarce in England where they really occur only on downland. Everywhere else is much too fertile and they are swamped by grass. In upland areas the soils are often thin, gritty and thoroughly unfertile, with the added influence of high rainfall that literally washes out any fertility enabling harebells to compete more easily.

There is an important point if you are trying to establish yellow rattle. Because it is an annual, it seeds and then dies, so it is advisable to mow or strim above it. Some tops are inevitably cut off, but by cutting at 15-20 cm (6-9 in), some seed sets and light and moisture are let in. Yellow rattle seed lasts in the soil several years, so it is not a complete disaster if all the seed heads are cut off but if the meadow is on a fertile site, getting yellow rattle to produce seed in the first year is important.

When you examine the amazing growth in your meadow remember that you have a not-to-be-repeated opportunity to discover rare cornfield annuals. You need a sharp eye because mowing or strimming are done before anything flowers, but mow round any unusual plants and allow them to flower so that you can identify them.

The mowing also benefits emerging perennials because some species such as ox-eye daisy grow much faster than, for example, cowslip or lady's bedstraw. If the ox-eye daisies are not kept in check, the establishment of the slower species suffers. For most people it is difficult to tell the difference between small weeds and small perennial wild flowers, so the safest course of action is to keep mowing and forgo the pleasures of wild flowers until the following year. As if all these weeds coming from the seed bank were not challenging enough, there may be others that cause particular problems.

I have mentioned dock seeds which live for ever in the soil. They are mown off like everything else but then have to be dealt with in the autumn or in the following year by spraying or digging them up. There is also the threat of Yorkshire fog, which is a rather coarse grass whose seeds lie around in ex-arable acid soils and on woodland sites. They behave like dock seeds in that they require low winter temperatures to break the seed dormancy and allow germination, so cultivating your land through the season will not unfortunately get these seeds to germinate. The good news about Yorkshire fog is that yellow rattle controls it well.

What if the seed does not germinate?

I have mentioned earlier my experiences of sowing a small field margin where through a variety of circumstances, germination was delayed for a full year, and I suggested that if germination is delayed, it may be wise not to write it off too soon.

Friars Court, Clanfield, Oxfordshire

A much larger scale version of delayed germination occurred in a 4 ha (10-acre) field in the Thames Valley not far from Faringdon. This was low lying ground near the Thames and prone to flooding so it had to be sown in the spring. The soils were alluvial with a fair amount of organic matter and the first that I knew that anything was amiss was when I received a telephone call in June from John Wilmer, the owner suggesting that I should call in because there did not seem to be much happening. This was an understatement. All that was visible was a few creeping thistle and some timothy that had come from dormant seed. However, the alluvial soil presented a curious sight. It had completely lost its structure, now being akin to that breakfast cereal known as grapenuts. You could scrape off the top 7 cm (3 in) with your hand. No moisture could possibly be retained so nothing had germinated. When it had been worked down to a good seedbed, and great care was taken over this, there must have been a spell of extreme heat. A "don't panic" strategy was agreed upon and subsequent events confirmed that this was correct. Most species germinated twelve months later.

Of the 18 species of wild flowers that were sown, 14 have established, with betony, meadowsweet, fleabane and devilsbit scabious yet to appear. These are all true wetland

Sequence of events in 4 ha (10 acre) field at Friars Court, Clanfield, Oxfordshire

Date of sowing: 14th April 2003

Germination in spring 2003: nothing except a few plants of timothy and some creeping thistle.

Autumn 2003: signs of seedlings.

Establishment 2004: 15 species of wild flowers and all the grasses.

Establishment 2006: 18 species of wild flowers:

birdsfoot trefoil	lesser knapweed	cowslip
field scabious	hedge bedstraw	lady's bedstraw
meadow buttercup	meadow vetchling	ox-eye daisy
ragged robin	ribwort plantain	self heal
common sorrel	tufted vetch	wild carrot
yarrow	yellow rattle	black medick

The cows on Hungerford Common have eaten off the first flush of grass and the meadow buttercups are growing back, but at half their height

One of our newly established wild flower fields at the farm after the sheep grazed it in May. Birdsfoot trefoil completely dominates the re-growth

species, so the conditions may have been too dry. In their place have appeared wild carrot, lady's bedstraw, field scabious and black medick. Because the field in effect missed a whole year's growth in 2003 there was plenty of bare ground for thistles to seed into, but John Wilmer sorted this out subsequently with a combination of topping and sheep grazing, and the thistles have now been eliminated. This conversion from arable to wild flower meadow is considered by Natural England to be one of the best examples in the area.

YEARS 2-6 – HIGH INITIAL FERTILITY

This is probably the stage when most difficulties occur. You have got safely through the first year. You begin to relax. Several things can now go wrong, but there is one important rule. The season's growth needs to be removed, all of it! This takes us back to the traditional hay meadow, where the hay was cut and removed and then the re-growth at the end of the season was grazed and by implication removed. Many of us are in situations where we do not have animals. Here are two suggestions, both probably unwelcome:

- if you top your meadow and do not remove the cut material, it will deteriorate rapidly.
- if you remove the proceeds of the hay cut and then just mow the re-growth without picking it up, the meadow will also gradually deteriorate.

But the timing of the hay cut at the end of July/early August is important. There are plenty of flowers and no one is in any hurry to cut them, especially as there seem to be butterflies enjoying the nectar. So nothing is done, the growth gradually collapses and the autumn grass then starts to push through from below smothering any newly shed seed that attempts to germinate. It even suppresses some of the less vigorous flowers that have established. It is finally cut in October or November. This goes against the first principle of the hay meadow, which is that the annual growth needs to be removed as near as possible to 25th July although this needs some qualification:

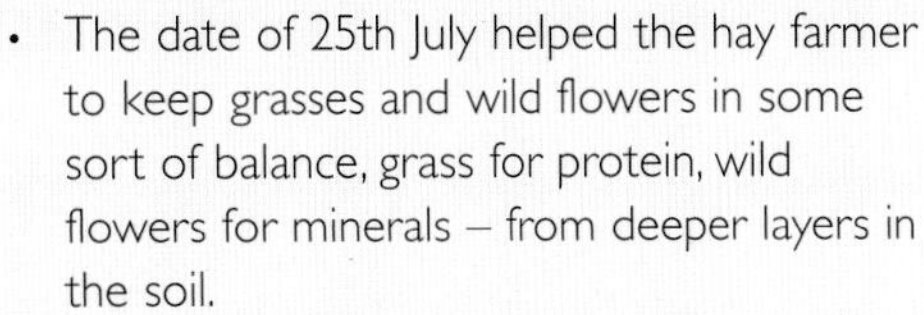

- The date of 25th July helped the hay farmer to keep grasses and wild flowers in some sort of balance, grass for protein, wild flowers for minerals – from deeper layers in the soil.
- Wet meadows with meadowsweet, dropwort, devilsbit scabious and great burnet would have been cut later, in August.
- In today's situation where there is almost no farm labour, getting someone to come and cut a meadow is increasingly difficult and may not be possible until the harvest is over.
- The nearer to the end of July that the meadow is cut, the greater the need for cutting or grazing the autumn re-growth in October.

The main point is that the annual growth has to be removed. The big question is how? Whether or not a complete job is made of cutting the hay, the area needs to be cleared of all the seasons's growth at the end of the year. When there is a warm autumn with abundant grass growth as in 2006, there will be a lot more grass to remove before the end of the year, and without grazing animals this can be difficult to achieve. Unlike the traditional farmer, we are interested only in the wild flowers. For most of us, grass is just a complication. I have seen countless meadows where there has been a build-up of grass in the bottom of the meadow over the years and wild flower diversity inevitably suffers.

We have found two methods of management to be successful:

Traditional hay meadow management

A. No action is taken in the spring, allowing everything to grow, with cowslips the first on the stage followed by all the other wild flowers. This ties in well with spring bulbs.

B. Cut the growth as low as possible in late July/early August and leave it to dry so that the seeds of the wild flowers fall out.

C. Collect everything up after a few days and remove.

D. When there has been significant re-growth, graze or cut, leaving the meadow as short as possible at the end of the growing season.

This sounds all very well, but what if the meadow is actually within your prize-winning garden, where a flock of sheep, however small, would be most unwelcome. A is obviously achievable; B might mean hiring an Allen scythe or bribing a local farmer to come round if there is tractor access; C depends on

opposite: The results of early cutting or grazing are seen at their simplest in many a lawn or close grazed field. Stop mowing or grazing in May or June and you will have a carpet of daisies

St James the Less Churchyard at Winterbourne, where selected areas are mown through May and then flower well into September. Lesser knapweed has very good nectar and benefits many insects, including butterflies, which in turn benefit the resident bats

manpower if the area is too small or too difficult for agricultural machinery, so family and friends may be important here. If all the season's growth is to be removed in one go, there is a lot of work to be done; D is where we begin to run into difficulties. The ride-on-mower is not the same as a grazing animal. Pulling the grass apart and treading in the newly shed seed is beyond it. This last action takes on great importance if we want yellow rattle to control the grass growth. Getting the main hay cut out of the way in late July or early August means that the seeds should be down near the soil surface by the time the grass has grown and is ready to be cut. The occasional dew will have added moisture and dampened the seeds and made them less prone to be hoovered up by the rotary mower. So what are our substitutes for cutting and grazing?

Cutting
When the ride-on-mower cuts the re-growth, the grass must be collected or a thatch of cuttings builds up. This may not occur in a single year, but over a few years it will prevent seeds, particularly the saucer-shaped yellow rattle seed coming into contact with the soil.

Grazing
As far as I know, there is no machinery that can mimic grazing. Close mowing and scarifying is all that is on offer, but grass harrows would also be effective. On a small scale, vigorous raking works well – another task for family and friends. The curious and maddening feature of yellow rattle is that it refuses to co-operate in circumstances that appear ideal, and yet works brilliantly when you least expect it to.

Early cutting and grazing
This is a wonderfully simple system. It avoids making hay or anything like it, and it almost keeps the task within the capabilities of the ride-on mower, well almost!

A. Graze or mow (2-3 cuts) until end May.

B. No further action until early September.

C. Cut the growth as low as possible in September/October and remove.

D. Final mow with cuttings collected November/December.

Regarding the early cutting, it is an advantage if the cut material can be collected but we have found that topping on a field scale without collecting has not caused any problems. This could be because biological activity is high in early summer, whereas if cut material lies around in the autumn, there is less biological activity to disperse it. In any case, the cost of collection is simply too high. B is achievable. Unless the ground is very fertile, C should be within the capabilities of the average ride-on mower. What is significant is that the early mowing or grazing will have already removed half the season's growth, so the task is greatly reduced as compared with the traditional hay meadow method, where all the season's growth has to be removed in one go. D might be unnecessary if there is a dry autumn.

Although early cutting or grazing may not suit spring bulbs, it is a system widely used where a hay cut is not required. On downland, sheep are put in to graze for 6-8 weeks when the cowslips are in full flower and having removed the peak of the grass growth, the way is clear for a great show of wild flowers later in the summer. On Hungerford Common near the farm, cattle graze the grass early in the summer with an immediate response from the meadow buttercups. This is no different from a great show of daisies if grazing has been hard. It is almost impossible to overgraze, but it is all too easy to undergraze. Mid-season grazing will, of course, remove yellow rattle, but in any case the grazing action is doing the job of yellow rattle by keeping the grass under control.

There is another major advantage if the early cutting regime can be run alongside the traditional hay meadow regime. Early cutting or grazing until early May puts back flowering 4-6 weeks. Thus, if the traditional hay meadow regime achieves flowering from mid-May to mid-July, early cutting results in a later start to flowering but it should continue until early September. In terms of availability of nectar, this is a massive boost for insects and their associated wildlife.

You may need to change from one regime to the other. In Cowfield, where we did not put yellow rattle in the seed mix, there was enormous growth in the first five years, so we took repeated hay cuts. When growth began to reduce, we switched to early grazing, which we still do today. There are many variations on removing the early grass growth. You can start mowing immediately after the cowslips have flowered and continue until early June.

Other management problems
Clover. This has emerged as a significant problem where ex-arable areas are converted to wild flowers because clover seed can sit around in the seed bank for a great many years. It can reappear and, if care is not taken, become dominant. As a precaution, it is not advisable to put clover seed in the seed mix. Red clover usually appears from the seed bank. If clover does appear, remember that it responds fastest to close cutting or grazing, so you need to do the opposite: not graze or not cut lower than 15-20 cm (6-8 in).

Spear thistle. This can appear in alarming quantities in the second year. There is a simple

Meadow Management Regimes

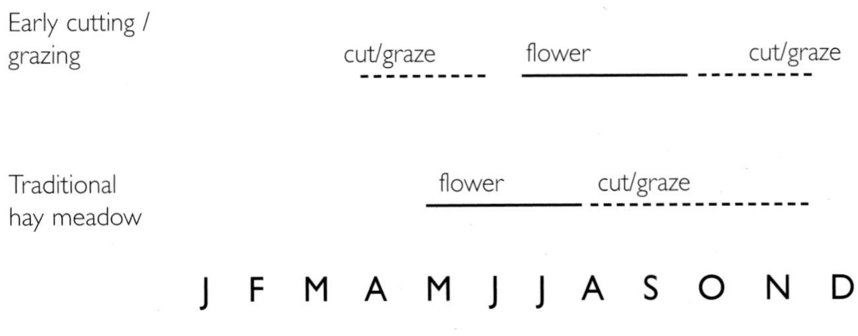

Early cutting / grazing — cut/graze - - - - flower ——— cut/graze - - - -

Traditional hay meadow — flower ——— cut/graze - - - - - - - -

J F M A M J J A S O N D

Early cutting / grazing delays flowering and, when combined with the traditional system, extends the flowering season for a further 4-6 weeks

solution: get out the mower just before flowering is about to start. You will lose some wild flower pleasure but the thistle will not flower or seed and it will not come again because it is a biennial.

Creeping thistle. This is a much more serious problem, because it not only seeds prolifically, it sends out underground rhizomes which quickly turn a couple of plants into a large clump. Creeping thistle has caused the destruction of many a wild flower meadow because the farmer had no other option than to spray a herbicide all over the field, knocking out the wild flowers as well. "What else was I to do?" I have heard many a farmer say. Apart from topping constantly, there was nothing he could do. Today, technology has moved on and we have the weed wiper. We used this with excellent results in our best wild flower field. Our sheep grazed down the vegetation in midsummer leaving the thistles ungrazed and then the weed wiper wiped a herbicide over everything that stood clear of the ground.

YEAR 7 ONWARDS

If your meadow is on agricultural land, it will have been pumped with fertiliser over the years, so the fertility levels will gradually drop. You may notice that the grass does not grow as high or that the wild flowers are doing better than they used to. If yellow rattle has become dominant, the grasses will have taken a hammering. But what you may see after seven or more years is that some wild flowers, which appeared to be only just managing to survive, are beginning to expand. We noticed this with some of the chalk species in our chalky field at the farm. Small scabious looked under pressure for the first seven or eight years. Now it is away. Dropwort is also increasing.

It is at this point that something may begin to happen that will be completely unexpected. As fertility drops, species start to find their way to your meadow. They may appear from the seedbank, move in from the hedge, or just drop in as wind-blown seed. It begins to get exciting.

Red clover is one of those species that seems to lie dormant in the soil for many years. Its appearance is particularly good news for bumble-bees. *Fairy flax* is an annual, which although a dainty little flower with a very small seed, also appears to lie around for ever in the soil, and it thrives if the area is hard grazed, for

Red clover

Fairy flax

Greater broomrape

Pyramidal orchid

example by rabbits. *Orchids* are a great puzzle to most of us because the seeds really are like dust. We are told these seeds get taken up by thermals and then get dropped by the rain. I am quite prepared to believe it! *Greater broomrape* is a parasite on knapweed, so it not only has to reach you but also find its host. Many other species may appear, and the history of the site clearly plays a large part in this.

SUMMARY

Management can be divided into three phases:

- Year 1. Establishment: repeated cutting or grazing overcomes any competition from annuals or delayed germination.
- Years 2–6. High initial fertility: removing the annual growth, traditional hay meadow management / early cutting and grazing, extending the supply of nectar.

  Reduced wild flower diversity is nearly always caused by under grazing or lack of management rather than the opposite.
- Year 7 onwards. Species respond as fertility reduces.

above: Unexpected arrivals

opposite: As fertility dropped, small scabious began to increase

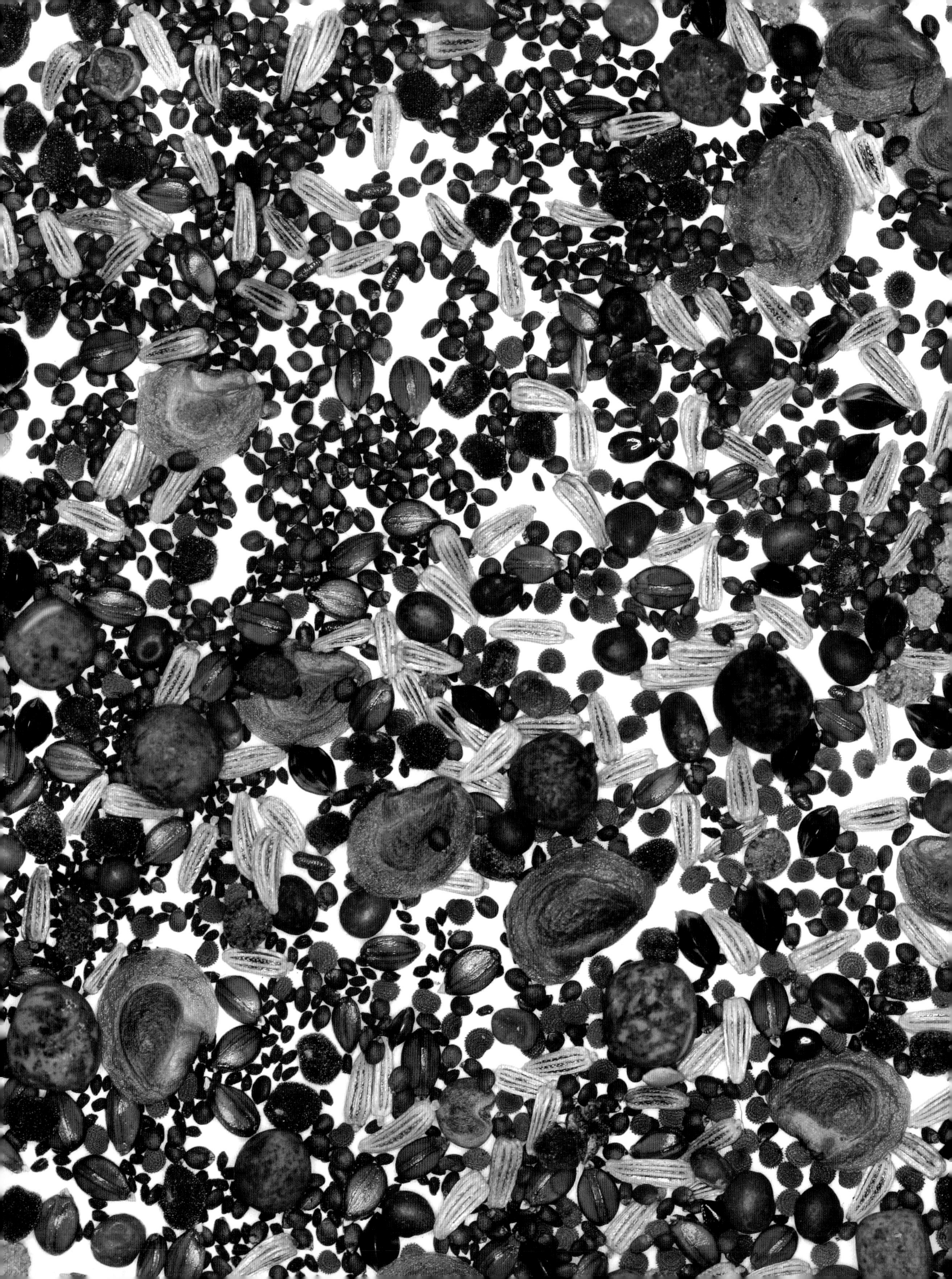

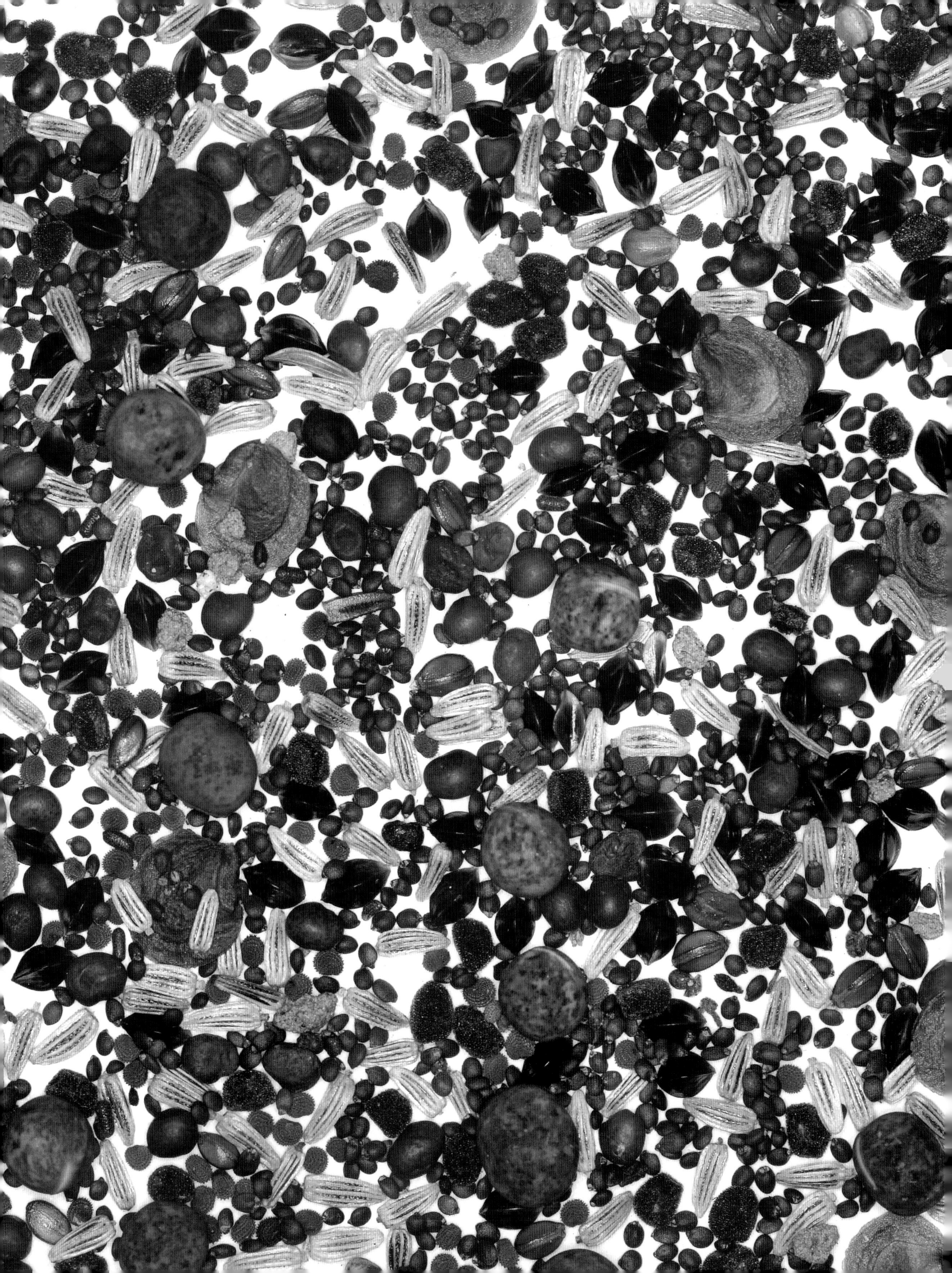

CHAPTER 9
PROPAGATING WILD FLOWERS

WILD FLOWER SEED IS DIFFERENT. WHEN YOU BUY A PACKET OF WILD FLOWER SEED YOU COULD BE FORGIVEN FOR THINKING THAT IT IS NOT UNLIKE ANY OTHER SEED YOU MIGHT BUY IN A PACKET, SUCH AS GRASS SEED OR VEGETABLE SEED.

But you would be totally wrong. There is one essential difference: the seed has not been through the hands of the plant breeder where such characteristics as quick and even germination and quick establishment have been carefully encouraged.

Wild seeds behave differently in that one of the key survival mechanisms is uneven germination. This was demonstrated to perfection in one of our fields when we sowed a grass/wild flower seed mix. Everything germinated well, even the cowslips, which are notoriously slow to get going. But we then had a hot summer and by the end of it all the cowslips had died. The following year the same thing happened. It was only in year three that we had a cooler summer and by the end of the season there was a scatter of young cowslip plants, which have gradually spread all over the field. If the cowslips had all germinated in the first year, we would have lost the lot.

Plants that grow in the key habitats of meadow, pond and wood have developed propagation strategies to suit their surroundings. In a meadow, seed has to fall to the ground in readiness for the autumn rains to soften the ground, so that the seedling's roots can penetrate the soil surface. In a pond, the seeds may float away and get blown to the shore where the competition is intense. Maybe this is why purple loosestrife needs to produce quite so much seed even though it has a seed size of 15,000 seeds per gram. However, many plants here have developed a rhizomatous strategy where roots can penetrate the soft soil and colonise underground. In woodland, where shade is a feature, plants have developed runners to colonise new territory because the light conditions are often insufficient to set seed. As for annuals, since they have only one chance to seed before they die, you would expect them to produce a lot of seed, but this is not always the case, whereas biennials such as foxgloves,

Cowslip seedlings are very vulnerable to a hot summer

Most seeds, such as ox-eye daisy, germinate in a few days

opposite: When cowslip seed has turned brown and hardened germination becomes erratic and can be delayed for several years – an important survival characteristic

pages 110-111: Yellow rattle seed, which is large and flat, finds it difficult to make contact with the soil. Contrast this with the smaller, round seeds, which can easily roll down to the soil surface

above: Kidney vetch is a good coloniser of the chalk and one of the first seeds to germinate

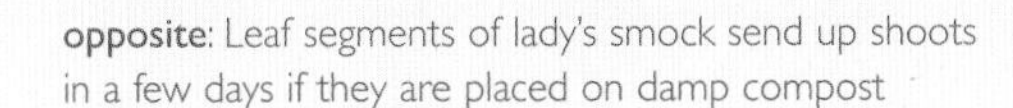

opposite: Leaf segments of lady's smock send up shoots in a few days if they are placed on damp compost

Lady's smock, a characteristic plant of damp grassland

which also have only a single seeding opportunity, produce far greater amounts of seed.

MEADOW PLANTS

When you spread your seemingly inert seeds in a seed tray, water them, and perhaps place some glass or polythene over the top to stop them drying out, small green shoots appear after anything from a few days to several weeks. Most of the commoner meadow plants such as birdsfoot trefoil, lady's bedstraw, common sorrel, knapweed, ox-eye daisy, self heal and yarrow germinate easily, but in some cases nothing at all may happen. This is due to the built-in survival characteristics of these seeds, which we need to understand. I have listed below some of the techniques necessary to get round germination problems, as well as other characteristics that are simply of interest to anyone growing plants from seed.

Winter chill. The primula family, which includes cowslip and primrose, have very hard seeds that are difficult to germinate. When they are planted in the autumn, nothing happens until the spring because a period of winter chill is required to break down the hard seed coat. When the seeds of a cowslip have been formed but before they are ripe, they are green and will germinate easily. As they ripen, they turn brown and become increasingly hard to germinate. This survival characteristic is a protection against all the seeds germinating together and then being wiped out by some catastrophic spell of weather.

Many books advise mixing cowslip seed in moist sand and putting it in a fridge, but the subsequent sowing is a nightmare with damp seeds sticking stubbornly to your fingers. It is much easier to sow the seed tray in the autumn and put it outside with some fleece over it for protection. It will then germinate in the spring when the temperature rises and days lengthen, the latter being an important growth trigger. Other species in this category include common rockrose and agrimony.

Scarifying. Certain plants have a tough seed coat that just needs to be roughed up a bit to let in the moisture. If you take some sandpaper to meadow cranesbill, it germinates much better. Soaking the seed for twenty-four hours also helps.

Covering seeds. With most of the smaller seeds, it is unnecessary to cover them with soil after sowing.

Larger seeds such as tufted vetch (80 seeds per gram) or meadow vetchling (100 seeds per gram) can be lightly covered. But there are awkward customers here. Field scabious seed germinates best when uncovered, but when it begins to grow, the shoot slithers along the soil surface and often seems loath to grow into the soil, so it is best to cover it once it has begun to germinate.

Autumn sowing. This definitely suits some species, particularly red campion,

Short-lived perennials. Some meadow plants usually classed as perennial are really short-lived perennials in that they may not live for more than a year. Ragged robin is one of these and requires open damp ground to really perform.

I have an unforgettable sight captured on an old photograph of the edge of a streamside meadow in our village where the previous

farmer's cattle used to congregate to look over the fence. One year the whole of this area, which was about the size of a tennis court, was scarlet with ragged robin. Self heal is another species that often behaves like an annual.

Good colonisers. Some plants are known for their colonising ability and it is always nice when the seeds really do live up to their reputation. Kidney vetch has a fairly bold seed (350 seeds per gram) and is in every seedman's calcareous seed mix. It is always one of the first wild flowers to germinate.

There is only one reasonably common species as far as I am aware which has developed an alternative strategy to growing from seed and that is to grow from leaf segments. Although it still produces a little seed, lady's smock uses its leaves to produce new plants. Leaves get detached from the parent plant, usually at the end of the winter, and then if they lie about in damp conditions, rootlets appear from the centre of the leaf. If the leaves are collected and placed on damp compost they will be growing within a week.

When growing seeds in seed trays all the usual rules apply. Seed trays must be cleaned before they are used again and composts should be reasonably fine to accommodate the small seeds of wild flowers. There is one feature of the seed tray that gives rise to a problem afflicting at least two of the commoner flowers: wild marjoram and lady's bedstraw. This is the "damping off" problem, where a fungus gets into the seedlings and can quickly spread to the whole tray. Most people fill their seed trays with compost to within a couple of centimetres (three-quarters of an inch) of the top, so that the glass or polythene placed over the tray remains clear of compost. It is in this space between compost and polythene, where there is no air movement, that damping off disease can run riot. As a result of this, for these vulnerable species we now usually use plug trays, which are filled up to the top and left

Ragged robin, which often behaves like an annual, requires grass growth to be removed by the end of the year leaving bare ground to help it grow successfully. It is seen here in open, pebbly ground

opposite: Close-up of ragged robin

Water mint is one of the fastest colonisers and can become over dominant

Pond sedge rhizomes will spread widely given waterlogged soil

uncovered so that there is sufficient air movement to avoid the disease.

AQUATIC PLANTS

Aquatic plants rely mainly on two strategies: setting seed in the usual way; and using root rhizomes. But many species that use root rhizomes produce seed as well.

Setting seed

There are two zones where seed has different characteristics. Marsh marigold needs water-logged ground and hence grows at the water's edge. Since it is not very competitive, growing at the water's edge is also of benefit. Its seeds do not germinate unless they are in water-logged soil, and a seed tray needs to be placed in shallow water.

Most plants that like damp soils grow in the capillary zone on the edge of the pond where moisture is drawn up from the water's edge. These are the wet meadow plants such as devilsbit scabious, great burnet, meadowsweet, common valerian and water avens. Their seeds are sown in a seed tray in the usual way.

Rhizomatous plants

Marsh plants are the main constituents of this group, which includes many sedges, branched bur-reed, greater spearwort, gipsywort, sneezewort, fleabane and meadow rue.

Some plants such as water mint and brooklime use surface roots almost like runners. These two species illustrate well the way each plant has a strategy for colonisation. Water mint colonises anywhere and is a powerful customer. Brooklime, on the other hand, can be easily outgunned, but its chance comes in the autumn when the pond dries and mud is available. It then produces at great speed runners that can cover considerable expanses of this mud. When the water rises again, it has to retreat.

The easiest way to propagate these species is to have a stock pond where you can simply dig up and divide the rhizomes as required, and then press them into the wet soil in the new pond. Some of them, such as the sedges, bur-reed, sneezewort and fleabane also produce seed, in which case it is sown in the usual way.

Features such as winter chill affect yellow flag iris. This has some of the larger seeds at 20 seeds per gram, but they are difficult to germinate. One spring a few years ago, we picked up a pile of seeds that had spent the winter on a pond dipping platform. They all germinated at once.

WOODLAND PLANTS

The key feature of woodland is shade. Shade limits grass growth and thus can give woodland plants a competitive advantage. But shade also reduces a plant's ability to flower and set seed. This has led woodland plants to use three strategies to colonise: seed, runners, and underground rhizomes

Primrose sets seed but has a curious way of tucking the seed heads under its leaves after it has flowered. The seeds are then removed by ants, which no doubt helps to spread them but makes life difficult if you are trying to collect them. Then, in the deepest shade imaginable there are carpets of bluebells, also with seed. The bluebell can flower in shady conditions, but if it is to set seed, more light is required.

Runners

Runners are the principal response to insufficient light. Species such as bugle, sweet woodruff, yellow archangel, barren strawberry and sweet violet send out runners. A single plant of bugle makes a large clump by the end of the growing season.

Runners should be dug up from a stock bed in spring, divided between bud joints, and planted in a small pot.

opposite: Marsh marigold has to survive at the water's edge in often very competitive surroundings. Note the rabbit wire to keep the muntjac away

above: Great mullein appearing from the seed bank on a set-aside field

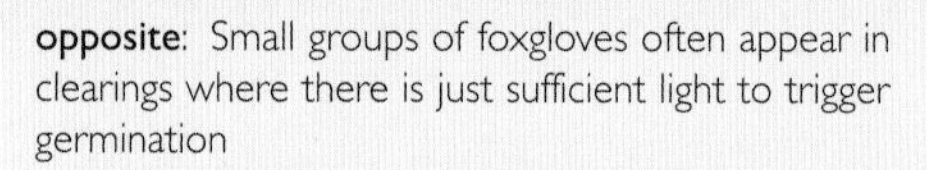

opposite: Small groups of foxgloves often appear in clearings where there is just sufficient light to trigger germination

Rhizomes

A number of species adopt this survival strategy. Lily-of-the-valley sends out strong rhizomes in all directions and really is quite aggressive in good soils, whereas Solomon's seal uses the same technique but almost in slow motion. Nothing much happens in a single growing season! Once again, the stock bed has to provide the plants, which are divided up in spring. Wood anemone has curious short rhizomes, which grow very slowly and appear rather like bent matchsticks. They are best divided or broken into shorter lengths in the autumn.

Annuals

It is easy to forget about this group of plants or just assume that most plants are perennials. People frequently ask for poppies to be included in their meadow seed mix assuming them to be perennials. Annuals need the ground to be cultivated for their seeds to grow. When a cornfield is ploughed and cultivated, a whole mass of weeds begins to grow. It is the same in a kitchen garden. This is a particular characteristic of annual and biennial seed. It may have to survive a very long time in the soil. There are stories of parts of the Berkshire downs being ploughed up in the Second World War for the first time since the Napoleonic wars over a hundred years earlier and the fields coming up scarlet with poppies in the next growing season. The seed had lasted over a hundred years. If you examine a poppy seed you will find that it is very hard (needing winter chill to germinate), small and round, in fact perfectly adapted to survive for a long time in the soil.

Biennials

Biennials pose particular problems in that they grow in the first year and flower in the second. However, they are not common in the aquatic environment although the woodland habitat has several common species, foxglove being the most loved. Fortunately, it produces huge amounts of seed and is not threatened in any way. Its seeds can easily survive a seventy-year rotation of fir trees. Other woodland biennials include marsh thistle.

top: Bugle colonises well from runners

above: Wood anemone grows slowly from curious, bent rhizomes but once established it is one of the glories of ancient woodland

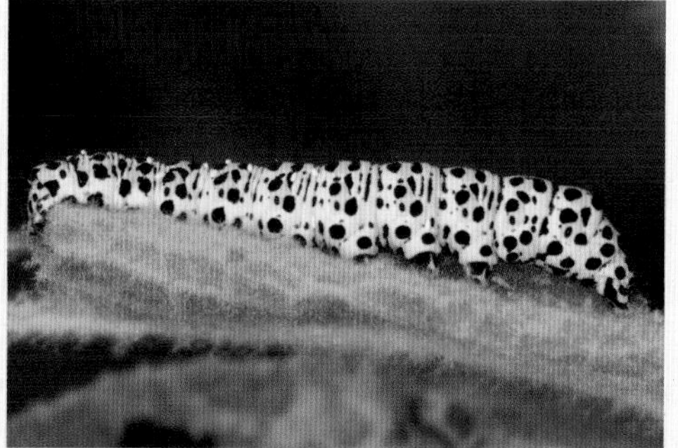

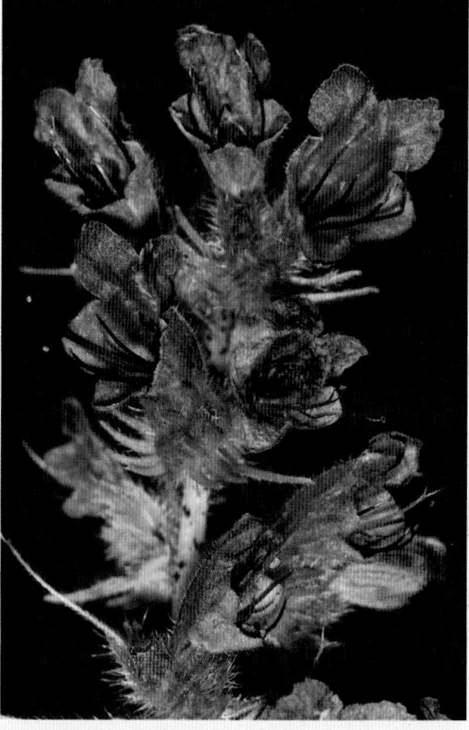

above: Dark mullein is an attractive plant that is widespread in chalky areas

top right: The mullein moth caterpillar can demolish all the leaves on a great mullein in a few days

centre: Teasels used to grow in so-called waste places, which have now almost ceased to exist, so we shall have to grow them in our gardens

above: Close-up of viper's bugloss

above: When you see viper's bugloss in full flower, covered in insects, it makes you want to champion its existence

It is the arable or wasteground habitat where many biennials exist and these have a most precarious existence. They include teasel, much loved by seed-eating birds and flower arrangers, viper's bugloss, which has valuable nectar for insects, great mullein, a plant that somehow survives in field margins, and plants such as the common spear thistle and some of the other less well-known ones such as welted thistle and musk thistle. Teasel and viper's bugloss have about 250-300 seeds per gram but they are not particularly well adapted to survival in the soil. Great mullein – and indeed the other mulleins, such as the dark mullein – has over 10,000 seeds per gram and once again the seeds are small, round and tough, and thus well adapted for survival.

In our area of East Wiltshire, viper's bugloss seems to be very scarce and given today's intensity of land-use, it makes me wonder how it will survive.

SUMMARY

Wild flower seed has "wild" characteristics such as uneven germination and has developed propagation strategies to suit its surroundings.

- Meadow plants usually germinate most easily in the autumn. However, some require winter chill or scarifying to aid germination. Seed trays attract fungal attack unless precautions are taken

- aquatic plants use seed and underground rhizomes

- woodland plants use seed, runners and underground rhizomes

- annuals and biennials have special characteristics which suit their opportunist life style

CHAPTER 10

SAVING OUR CORNFIELD ANNUALS

THE POPPY IS A MEMBER OF A GROUP OF PLANTS DESCRIBED AS CORNFIELD ANNUALS, WHICH ARE SOME OF THE MOST THREATENED PLANTS IN BRITAIN.

There are two factors that have brought this about. Most cornfield annuals do not have such tough seeds as the various species of poppy; and secondly, since the arrival of herbicides, a single pass of the farm sprayer can wipe out a generation of these cornfield weeds. It is this second point that is the most serious. If arable crops are grown year after year, and the farm sprayer then destroys the arable weeds year after year, the seed bank of cornfield annuals will gradually be reduced until there are none left. So why are poppies still so common? The answer lies in the fact that poppy seed is around in prodigious quantities. For example common poppy has 10,000 seeds per gram, whereas corn marigold has 600, cornflower 200, corn buttercup 100, and corn cockle only 60. Those species with smaller numbers of seeds per gram will become extinct first, unless of course the appropriate conservation measures can be put in place. Corn cockle is now extinct in the wild in Britain. Some of these cornfield annuals have developed so closely with corn crops that they either germinate in spring with a spring sown crop or in autumn with an autumn sown crop. Corn marigold germinates in spring barley (it is still not uncommon in parts of Scotland), and corn buttercup germinates in autumn sown crops.

It came as something of a surprise to find that we had a field where there was a small population of corn buttercup, but our efforts to conserve it demonstrate just how difficult it is to get the management right.

CORN BUTTERCUP (*Ranunculus arvensis*)
It was a stroke of good fortune that Phil Wilson who happened to be studying Britain's arable flora appeared at the farm at the precise moment that I was trying to convert the farm to an organic system (unsuccessfully as it turned out). Our first field to be converted had been put down to grass and clover, but too late in the autumn. By the following spring, it was a thoroughly poor crop. However, in the gaps were a fair number of corn buttercups, a nationally scarce plant, which Phil spotted during his visit. Otherwise we would never have known they were there. The field of grass and clover was duly made into silage and the buttercups were unable to flower and set seed.

This attempt to go organic failed for a variety of reasons and the field was soon growing conventional corn again. It was fairly easy to see the distinctive young corn buttercups when the crop was only a few inches high, but as soon as the crop was fully grown it was simply too dense for the buttercups to survive.

We attempted to clear around one or two plants to see if we could get them to flower and set seed, but it was difficult and unsatisfactory. A few years later we cultivated the bottom two acres of the field and left it unsown. We were rewarded with a magnificent field of weeds but as many as 80 plants of the corn buttercup, which flowered merrily and seeded.

Cornfield annuals en route to extinction: corn cockle (above) is now extinct in the wild; cornflower (top) is now on the nationally rare list

opposite: Corn marigold is still widespread

pages 124-125: The beautiful cornflower is approaching extinction in the wild but it can easily be reintroduced

Seedling corn buttercups before they were swamped by the well-fertilised corn crop

above: The corn buttercup seed is covered in an extraordinary set of spikes

opposite: Corn buttercups are adapted to fight for survival in a corn crop

What were we to do next? We should have ploughed the field, but we did not. Instead, it was cultivated, which did nothing to check the creeping thistle that had appeared. The following year we had fewer corn buttercup, a mass of weeds, and a fine crop of creeping thistle. The field looked a real mess! It took a lot of effort and herbicides to clean it up.

In due course the farm was converted to an organic system, and I was relishing the thought of corn buttercup growing in a not too dense corn crop. Alas, the grass and red clover we had grown to build the fertility for our first organic crop of wheat had been so effective that the wheat was almost as thick as a conventional crop. The corn buttercup did not stand a chance, so we had failed again.

However, with the new Environmental Stewardship Scheme we are trying again, with a 0.8 h (2 acre) cultivated strip along the bottom of the field where we know there are buttercups. This time we shall move the strip up the field each year to prevent a weed build up. A reasonable number of corn buttercup flowered and set seed in 2007. They were not particularly large plants, but each probably had 10 seeds and we had at least 100 plants, so I think at last we are making progress.

MOUSETAIL (*Myosurus minimus*)
This plant is described in the Wiltshire Flora as "nationally scarce". It was observed in 1985 not far from the farm, which was the first record since 1957. There were also records of it in Wiltshire in 1988 and 1991. Our experience demonstrates that it may well be lying around in the soil in prodigious quantities, but this is impossible to verify.

A single but very robust plant was observed by a botanist friend in a new wild flower / grass seed mix sown in Pond field in 1994. The following year an adjoining small piece of woodland was extended by 1.2 h (3 acres): the stubbles were sown to a grass cover crop and the trees planted the following spring. The ground cover was still a bit patchy and in the gaps in the grass were hundreds of mousetail. We estimated that there were approximately 5 plants per sq.m over part of the three acres, or a conservative estimate of 10,000 plants. Of course, its real status can never be known except in one-off situations such as ours when perfect conditions were created for it to appear. In addition, it is not exactly the most exciting plant to look at and is easily missed. In fact, most plants, unless they are colourful like cornflower, are likely to be overlooked.

TRACKING DOWN CORNFIELD ANNUALS
At present much effort is being put into an attempt to discover the status of the cornfield annuals in the North Wessex Area of Outstanding Natural Beauty. It demonstrates just how difficult this kind of survey is. Areas where cornfield annuals have been found in the past can be identified, but what do you do next? Field margins of likely fields can be walked at the optimum time of year and gateways examined, and plants may be discovered but the discoveries at our farm were all within the crop. I have no doubt that discoveries will be made but this is like searching for the proverbial needle in a haystack.

THE CORNFIELD ANNUAL SEED MIX
Although many of our more colourful cornfield annuals are on the danger list, we can still enjoy them by sowing a cornfield annual seed mix consisting of such species as corn cockle, cornflower, corn marigold, common poppy and corn chamomile. Sowing can be in the autumn if your soil is free draining, or in the spring if the ground lies cold and is in a frost pocket such as our farm at Shalbourne. If these annuals are sown in April or May you will have a fine show of colour 2-3 months later.

How long can these plants be kept going without re-seeding? If the ground is clean, they should be cut to the ground after flowering and then the ground raked well. If there is any weed, use a herbicide before there is any risk of the seeds beginning to chit. It is best to sow again after two years, since there is an inevitable weed build up. In addition, some species such as corn cockle produce much more seed than, for example, cornflower, so the variety of colours will diminish. There are endless opportunities for experimentation.

SUMMARY

Cornfield annuals are some of the most threatened plants in Britain.

- Corn cockle is already extinct in the wild and other species will follow unless conservation measures can be taken.
- Our experiments with corn buttercup (nationally scarce) have demonstrated that it is not possible to grow cornfield annuals in a commercial crop. However, leaving an unsown strip without a crop has proved successful.
- The cornfield annual seed mix is colourful and can be producing a mass of colour within two to three months.

opposite: The only seed mix that can give colour the same season it is sown

top: Many customers ask for poppies in their meadow. Alas, common poppy is an annual and cultivation has to occur annually

above: Cornflower has spread into this seed crop of corn marigold. This sight could not occur in the wild because cornflower grows in autumn-sown corn crops and corn marigold in spring-sown corn crops

Roadside hedges are particularly good for wild flowers because when they are regularly cut back, the verge is trimmed at the same time

CHAPTER 11
RESTORING WILD FLOWERS TO THE HEDGEROW

OST OF US HAVE AN IMAGE OF A FAVOURITE HEDGE, REAL OR IMAGINARY. THERE ARE MANY DIFFERENT TYPES.

The key difference is whether the hedge is tall, wide and unmanaged – in which case the shrubs will have shaded out the wild flowers – or whether it is a managed hedge with plenty of wild flowers and butterflies, and with yellow-hammers singing from the occasional song post. If the objective is maximum diversity then it is the latter that is required. It is an unavoidable fact that maximum diversity goes with maximum management. If wild flowers are to grow along a hedge, the establishment and management of the hedge is crucial and so a digression is now necessary.

PREPARING THE SITE

If weeds such as grass, nettles or creeping thistle get a hold during the first two years of the hedge, you might as well have saved yourself the trouble and not bothered to plant. The hedge will not get away. Here are some rules:
Clean stubbles: no problem. Get planting.
Grass weeds: spray with a herbicide before the end of September. If couch is well established, ploughing will not achieve much as it will grow again from its rhizomes.
Nettles/docks: spray with a specific herbicide the previous May and again in autumn.
Creeping thistle: spray with specific herbicide in previous May.
If the hedge line is less than 100 yards and there are strong men about, chemicals could be avoided but getting rid of creeping thistle would still be difficult. Rotovating the hedge line also helps (best done before the end of September) but is not essential.

This overgrown hedge looks great in the landscape but wild flowers will hardly feature

pages 132-133: Winter is not over until the blackthorn flowers; hence the term "blackthorn winter"

SELECTING THE SHRUBS

When parliamentary enclosures were at their peak in the eighteenth and nineteenth centuries, some 200,000 miles of hawthorn hedge were planted. They were known as quickthorn rather than hawthorn because the hedge got away quickly, which was essential to break up the medieval open fields and make the new fields stock proof. Barbed wire did not appear until 150 years later. These hedges are single species hedgerows, whereas a thousand year old hedge could have as many as ten shrubs in it.

Max Hooper developed a theory in the 1960s that a 30m (100 ft) hedge accumulated one additional woody shrub for every 100 years of its existence. Today we have the opportunity to approach the subject from the other end and ask how many shrubs we would like in our hedges. We have a list of ten species so we have settled for a thousand year old hedge.

%	Shrub	Flowering time
15	hazel	Feb/March
8	blackthorn	March
5	field maple	April
2.5	wayfaring tree	May
2.5	holly	May
50	hawthorn	mid May
5	buckthorn	May/June
7	dogwood	May/June
2.5	spindle	May/June
2.5	dog rose	June/July

Honeysuckle should also be on the list but there are so many local varieties that only local sources should be used. Guelder rose and goat willow have been omitted since they are not really suitable as hedging shrubs. Elder has also been omitted because birds bring it in all too easily and elder in a hedge is often deemed a sign of dereliction.

Flowering time

Although the different leaves of shrubs attract many different insects, it is the supply of nectar

Goat willow

Blackthorn

Holly

Dog rose

Hawthorn

Honeysuckle

above: A variety of shrubs provides nectar for half the year: goat willow in February, blackthorn in March, field maple in April, hawthorn in May, dogwood in June, dog rose in July

and its availability that is of most importance with all the different effects higher up the food chain. Thus, if hawthorn was the only species you would have flowers for a two-month period in May and June but with a ten-species hedge the nectar period is increased by a factor of three to half the year.

When planning a supply of nectar it is also worth considering two additional shrubs to extend the season still further. Goat willow as mentioned above is not ideal for a hedge, but it can be planted in field corners and allowed to grow to full height and produce the pussy willow flowers beloved of bumble-bees as early as February. And if you tolerate ivy in a hedge or have some old stumps, it will flower until the end of October. Ivy is used by the Holly Blue butterfly for egg laying (as is holly), another reason to value it. Two other butter-flies lay their eggs on hedging shrubs: the Brimstone on buckthorn, and the Brown Hairstreak on blackthorn.

Structure

There is more to planting a hedge than planting a variety of shrubs randomly because the hedge needs structure.

Hawthorn, even when only a few years old has firm upright growth and thorns, and gives a hedge its robust structure. Thus, if the hedge is planted as a double row with four plants per metre (3 ft), one row can be entirely thorn. One of the reasons for this is that some of the other hedging shrubs, such as dogwood are soft and present no sort of barrier for many years. If you had dogwood in both rows, deer or humans could push through easily. The following are some of the factors that should be taken into account when calculating shrub percentages.

Hazel 15%. The key species for cutting and laying with a good structure. It used to be wonderful for hazelnuts, but the universal autumn practice of "nutting" will be out of fashion as long as the grey squirrel (introduced at the end of the nineteenth century) dominates our countryside.

Blackthorn 8%. This shrub suckers freely and is valuable in creating random thorny thickets, which provide almost magpie proof nesting

Ivy flower

Primroses will spread along the north or east side of a hedge

sites. It is not so helpful when the hedge borders permanent grassland and the blackthorn creeps ever outward. It may be wise to reduce this percentage in these circumstances.

Dogwood 7%. It produces an excellent supply of nectar but its soft growth provides little structure for many years. It colours well in autumn and spreads by suckers on lighter soils.

Field maple 5%. The field maple available from the main nurseries appears to be different from our native field maple. It grows faster so it outgrows the rest of the hedge and because of this becomes brittle when being laid. On the plus side it colours up with a wonderful bright yellow in autumn, which is striking but not really what our native shrub should look like.

Buckthorn 5%. A very dependable hedging plant with a good structure.

There are then four more species which each contribute 2.5%. No hedge should be without a cascade of *dog rose* flowers. *Holly trees* are a feature in any hedge but should not be overdone and are costly to plant. *Spindle* has striking pink fruits in autumn but some authorities do not accept it as truly native although it has been around in the south of England for at least four hundred years. And finally, the *wayfaring tree* has flowers with high quality nectar that turn into striking red berries.

The structure and growth habits of two of these species have an interesting effect on the

Instant hedges of the enclosure period have only two or three shrubs in them, but land use changes over the years make it almost impossible to work out how wild flowers fared in these hedges

If the fence is placed too close to the new hedge, it will be difficult to cut back the hedge and maintain a margin of wild flowers

Today's field maple nursery stock grows very much faster than our native species and the wood is half the density

development of a hedge. Blackthorn and dogwood both produce suckers and if they are planted randomly on one side of a new hedge, the hedge will begin to grow out at certain points, creating perfect scalloping. The bays or indentations created by this growth provide excellent shelter for birds and insects, particularly if wild flowers are added in the bays.

However, for the wild flowers to persist, the scallops have to be managed, which could lead to much discussion with the man on the hedge trimmer.

Provenance

A substantial premium has to be paid if you want shrubs grown from local seed sources. In most of our counties local sources do not exist because they could not begin to compete. Does it matter? The main argument put forward by organisations such as Flora Locale that campaign for the use of native plants is that if hawthorn is sourced from Italy or if an Eastern European shrub comes into flower two weeks earlier or grows later into the autumn, it no longer coincides with the life cycle of all those insects that have evolved with it. I find that field maple is the most glaring example of a small tree that is definitely different and it is worrying that we are using all this foreign material when we have not the slightest idea of the long term effects it might have on our invertebrates.

PLANTING

When to plant

Although the planting season extends from the time the shrubs are lifted in the nurseries in late October to the end of March or even April, there is no doubt that planting earlier helps the plants to establish and increases their chances of surviving one of those bone dry periods that occasionally occur in April or May. If you buy plants in November, heel them in and dig them up in early January you will see that a mass of small roots has developed. They want to grow!

Size of shrubs

For a small run of hedge of 50 m (165 ft) or so, root size may seem academic, but if the run is over 500 m (1650 ft), fitting the plant size to the slot made by the planting spade is crucial. Plants with roots that are too big cause a huge amount of extra work. In any case, a smaller shrub usually establishes better so we always go for 40-60 cm (15-24 in) plants.

Planning the planting

Planting a hedge along 25 m (80 ft) of a garden boundary is a fairly relaxed affair. Planting a 500 m (1650 ft) hedge across a former cornfield with two thousand plants needs serious planning. The essential steps are:

- Planting often takes place in rain, so it is no good having planting plans because pieces of paper disintegrate in minutes. With one row 100% hawthorn, grab a bag of hawthorn and get planting. But what of the species in the other row, which are all in different bags? The answer is to pre-mix the shrubs in bags of, say, twenty-five. Then, however heavy the rain, the plants can just be pulled out of the bag and planted.

- Line up fence posts or tree stakes every 50m (165 ft) to get the hedge line straight.

- Use a line of 50m (165 ft) or 100m (1650 ft) along the ground as taut as possible (we use a length of binder twine with a knot every metre (3 ft).

- Open up a slot with a planting spade.

- Place a cane in the slot.

- Plant the row of hawthorn at 1 plant per 50 cm (20 in) with the cane. Tread in firmly.

- Wrap the spiral guard round plant and cane.

opposite: Goat willow is the first shrub to produce abundant nectar and it is therefore incredibly important

Matching the size of root to the planting spade is essential if a long run of hedge is to be planted

Planting with the cane

Wrapping the spiral guard round the plant and the cane

- Remove the line so that it can be set up for the next section.
- Plant the second row with a mixture of shrubs (pre-mixed in a bag) repeating the planting procedures.

Keep the plants in the shade

Another key point of even greater importance is the way plants are handled before planting. If the day is drizzly and damp, plants left in the open bare rooted cannot come to much harm but what if it is a glorious day with hot sun in late March? A bare-rooted plant will dry out in a couple of minutes. It must remain in the planting bag until the hole is dug with the planting spade. When it is removed from the bag it can be placed in the shade of the hole without being exposed to the sun for more than a second! The temptation to lay out plants when the sun is shining must be resisted.

Maintenance years one and two

Depending on how effectively you cleared the hedge line of weeds, a single visit early in the season with a knapsack of herbicide is the key action in the first year and again in the second. The spiral guard keeps the chemical off the new plants and if spraying is carried out sooner rather than later, the weeds in the bottom of the hedge will not be more than a few inches high. If the job is left until the weeds are 1.2 m (4 ft) high, you will have a problem.

How far apart should the rows be?

We have found about 1 m very effective. This gives enough space between rows to plant the hedgerow trees in protective tree shelters at an approximate rate of 5/6 per 100 m (330 ft).

However, there is a variation on this theme to create a wider hedge with two rows planted 1.5 m (5 ft) apart, with the third row in the centre made up of blackthorn and hawthorn.

ESTABLISHING WILD FLOWERS

There are two opportunities to do this.

Two years after planting

After two years, provided that the hedge line has been kept free of weeds, run a rotovator – the minimum width of most rotovators is 60 cm (24 in) – along each side of the hedge in late August/early September and sow a mix of grass and wild flowers.

Keep it strimmed or topped the following year.

What species should be sown? This brings us back to the extremely artificial nature of a hedge that is a cross between a highly organised strip of woodland and a very narrow piece of wild flower meadow. One of the key considerations is the orientation of the hedge.

East/west hedges. With one side facing north and the other side facing south, this gives the opportunity for the widest range of wild flowers: woodland to the north, meadow to the south (see the table opposite). The crucial importance of the flowers lies in the nectar they provide. The combination of woodland and meadow species lengthens availability. Meadow wild flowers do not appear until May or June whereas sweet violets, celandines and primroses begin flowering in March and are followed by species such as red campion and greater stitchwort, after which meadow flowers begin in May and continue right through to August with fleabane.

A single but crucial visit to the newly planted hedge is required in the spring after planting before the weeds grow too tall

If the nettles are hiding the new hedge, it is almost too late to spray

Results of effective herbicide treatment

North/south hedges. The east-facing side of the hedge will be best adapted to the woodland species with the meadow plants on the west-facing side. If trees have been planted in the hedge, woodland plants can be established beneath them.

Where new hedges have been established on grassland with wild flowers planted along them, the stock fence needs to be far enough away from the hedge so that the flail head can not only trim the side of the hedge but also get at the base to trim back grass/wild flower growth. If the fence is too close to the new hedge, management will be difficult.

After cutting and laying

The second opportunity to establish wild flowers arises when the hedge is being cut and laid for the first time some ten years after planting. A bare strip of ground emerges either side of the hedge where it has been cut back. This is an excellent area to establish wild flowers with a full growing season available to clear up any nettles or other weeds that try to invade. The selection of wild flower species and sowing is as above.

Table of useful wild flowers for the hedge

All locations	*Climbers*
Agrimony	Greater birdsfoot trefoil
Common St John's wort	Hedge bedstraw
Field scabious	Meadow vetchling
Fleabane	Tufted vetch
Hemp agrimony	
Lady's bedstraw	*Woodland plants*
Lesser knapweed	Foxglove
Meadow cranesbill	Greater celandine
Ox-eye daisy	Greater stitchwort
	Herb robert
Chalk only	Primrose
Dark mullein	Red campion
Greater knapweed	Sweet violet
Small scabious	Wood avens
Wild marjoram	

A small number of wild flowers have adapted so that they can use the hedge as a climbing support to help them compete more effectively. They are generally robust and tall enough to survive the minimal management that wild flowers in a hedge bottom receive.

Woodland plants such as violets and primroses can be low growing if the shade of the hedge is strong enough to keep at bay grasses that would otherwise out-compete them. If trees are planted in the hedge, they give additional shade and permit increased wild flower diversity. Biennials such as foxgloves need individual treatment and may only be practical in smaller schemes.

Wildflowers that can climb into a hedge

Hedge bedstraw

Tufted vetch

Meadow vetchling

above: musk mallow is one of the many wild flowers that will thrive beside a hedge

opposite: Greater birdsfoot trefoil

SUMMARY

The management of a hedge determines the extent of wild flower diversity.

- Effective preparation of the hedgeline before the hedge is planted is essential if wild flowers are to be established.
- A mix of shrubs will give nectar for half the year, which can be supplemented by the wild flowers.
- Care should be taken with shrubs that sucker out from the hedge.
- The new hedge should if possible be planted before the end of the year.
- A planting plan is helpful for a hedge over 50 m (165 ft) long.
- The use of herbicides in years one and two ensures that grass weeds do not stunt the growth of the new hedge.
- Wild flowers can be established two years after planting or when the hedge is cut and laid.
- Meadow flowers can be established on the south side of the hedge and woodland flowers to the north.

CHAPTER 12
RESTORING WILD FLOWERS TO PONDS

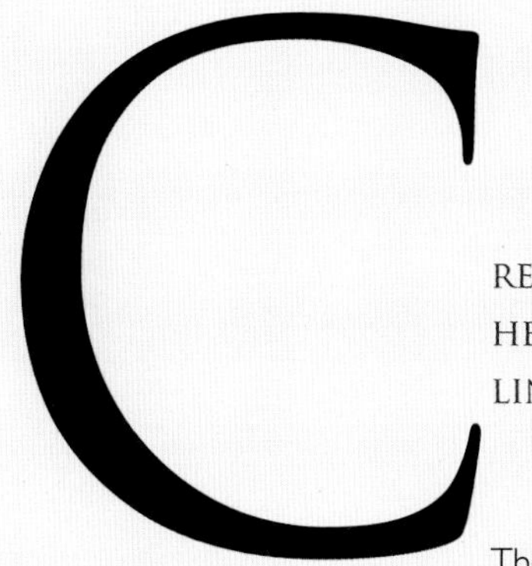

CREATING A POND IS MORE OF AN UNDERTAKING THAN PLANTING TREES OR A HEDGE. IT REQUIRES AN EXCAVATOR TO CREATE THE DEPRESSION AND A BUTYL LINER OR CLAY TO BE INSTALLED TO RETAIN THE WATER.

There is no doubt that for quick results the liner has to be the number one choice. If you start operations in May or June (assuming there has not been a really wet spring), install the liner, fill the pond and plant it up with a couple of dozen species, you will, within a few months, have attracted a variety of aquatic invertebrates such as water boatman and fresh water snails, at least half a dozen species of dragonflies, as well as newts, frogs and toads.

But in addition you will have created a watering hole for birds, mammals and anyone else in the wildlife world that needs to quench their thirst on a hot summer's day.

This is why a pond is so rewarding if it is in sight of the house or at least within telescope range, if only to watch the antics of the moorhens. But one of the great joys of an area of water is that you never know what is going to turn up.

The question of aquatic plants especially those with attractive flowers, is so dependent on how the pond is made and designed, that this now needs to be considered.

opposite: Betony is a plant of damp grassland and goes well by a pond

above: Damselfly dance. Common blue damselfly colonised Valley Meadow Pond as soon as it held water. They swarm in their hundreds

pages 144-145: Many species of dragonfly prospect a new pond even before the marginal plants are established. Ruddy darter is found in ponds, lakes and ditches and may even sit on your pen as you make notes

above: A butyl liner is the best material to keep water levels constant

Butyl liner or clay

There is a lot going for the butyl pond liner. It is easy to handle with a useful blanket-like material that accompanies it called "geotextile underlay" which is laid underneath the liner to protect it from sharp stones and over the top of the liner where the planting zone is planned. If the liner is left uncovered where the water is more than 30 cm (1 ft) deep, plants will not become established there and even if they attempt to do so, it will be easy enough to pull them up as their roots cannot attach to the butyl liner. The significance of the depth at which the liner can be left uncovered is that ultra-violet damage is greatly reduced once there is more than 30 cm (1 ft) of water.

An alternative to the liner is clay. Traditionally ponds were made using puddled clay with varying amounts of straw to help bind the clay, which was then trampled by sheep to close up any cracks. This kept the amount of clay to a minimum. Today we have run out of small flocks of sheep to do the trampling and have to rely on tracked excavators with rather more clay, perhaps as much as 50 cm (18 inches). The clay will almost certainly outlast the butyl liner but the logistics must be compatible with the site.

A lot of space is required to shift the small mountain of clay usually required. But there is another important point about the way the butyl liner reacts to dry ground conditions, which makes the management of pond margins more predictable. Because the liner protects the water from the evaporation pressures exerted all round the pond in a dry spell, plants are less likely to be left high and dry, although in a really hot spell, liners may need to be topped up.

above: Southern Hawker breeds readily in garden ponds. It is often seen away from water in woodland rides

What makes the clay-lined pond a little suspect is that as the surrounding ground dries, moisture is inevitably sucked out of the pond as the surface water evaporates. The situation of a clay-lined pond today could not be more different from that of two hundred years ago when our enclosed landscape was created. All they cared about then was a reasonable supply of water, not the attractive wild flowers around the edge. Flocks of sheep were regularly herded into ponds to give the margin, which might have been at risk from cracking, a good puddling. If there was a leak and the pond dried out, more chalk or clay and straw was laid on and the flock of sheep was herded into the pond to do the rest. The point is that there was a lot of management effort and skill to keep all those ponds in good shape, particularly dew ponds on downland. Under today's conditions, if the water level of a clay-lined pond drops and there is no means of topping it up, the clay cracks and there is a real problem. In addition, once water levels drop, aquatic vegetation quickly spreads across the pond and the water surface area diminishes.

POND DESIGN

Spoil from the excavations

The problem of where to put the spoil can be turned into an opportunity in that ponds need shelter from the cold north and east winds. Earth banks can be created with shrubs planted to increase the height of the protection and with rough grass as ground cover. This will provide ideal over-wintering habitat for amphibians such as newts, which rather sensibly leave the freezing cold water in winter and spend this period deep inside a warm tussock of grass. Suitable shrubs would include alder

Male Black-tailed skimmer. It is quick to colonise new ponds, aggressive, and fun to watch

buckthorn, dogwood, guelder rose and wild privet planted no closer than 4-5 m (12-15 ft) from the edge of the pond. Tufty grass can be achieved by sowing a red fescue at 5-10 grams per sq.m (0.2-0.35 oz per sq.ft).

Position of the pond in the landscape

This may be easier to think about when you have considered the above. Ponds that look right in the landscape are not too deep or high as compared to other ground levels. If there is some ground which can shed water into the pond so much the better. Remember also that you may wish to supply water to your pond from a nearby roof. A balance may have to be struck between not having the pond quite as deep as you would like because the amount of earth to be moved and subsequently "lost" would simply be too great. If the ground beneath the proposed pond is liable to springs in wet weather, you may require a drain under the pond to stop water from coming up from under the liner and causing it to balloon.

Providing a beach

This is all to do with providing an attractive drop-in point for passing birds such as wagtails and migrating waders. The dynamics of pond vegetation (and this is where your planting scheme comes in) quickly leads to a thickly vegetated pond margin. If you look at the average gravel pit, the bare, gravelly margin turns into a band of willow scrub in only a few years. The idea of the beach is to keep the vegetation at bay and the method to keep it this way is to put down some geotextile underlay along several metres of shoreline with a layer of 20 mm gravel on top. Any vegetation trying to invade can be removed and the beach will stay open for many years.

Planting areas

The simplest design for a pond is to make it like a saucer with shallow margins and a deeper centre of a minimum of 1.5m (5 ft). JCB drivers have now dug so many wildlife ponds that they will probably tell you exactly how it should be designed. Within this framework, there are two additional requirements: a wider shallow area of at least 1m (3 ft) in width; and a deeper shelf of approximately 20-30 cm (8-12 in) where deep water plants can be established.

Trees

If there is a tree that is too close to your proposed pond and will dump all its leaves into it every autumn, there is an alternative to felling. Pollarding is a marvellous compromise where you retain the main trunk with all its lichens, mosses and attractive bark, but remove the top. Trees respond to this treatment remarkably well.

PLANT GROUPS

There are three groups that need to be catered for in the pond design (see diagram):

Zone A – Water's edge, uncompetitive

These plants need space away from the more competitive plants and should be planted either side of the shingle beach or in damp grassland. Management requirements are to ensure that plants from Zone C do not invade, so Zone B plants can be used as a buffer against Zone C.

Devilsbit scabious	Lady's smock
Marsh marigold	Ragged robin
Common valerian	Brooklime
Water avens	Betony
Square-stalked St John's wort	

opposite: Surgery or pollarding will retain the tree as a feature but greatly reduce the shade and volume of leaves falling into the pond

pages 152-153: Bogbean grows out over water offering a useful protective covering to newts

Planting zones and liner

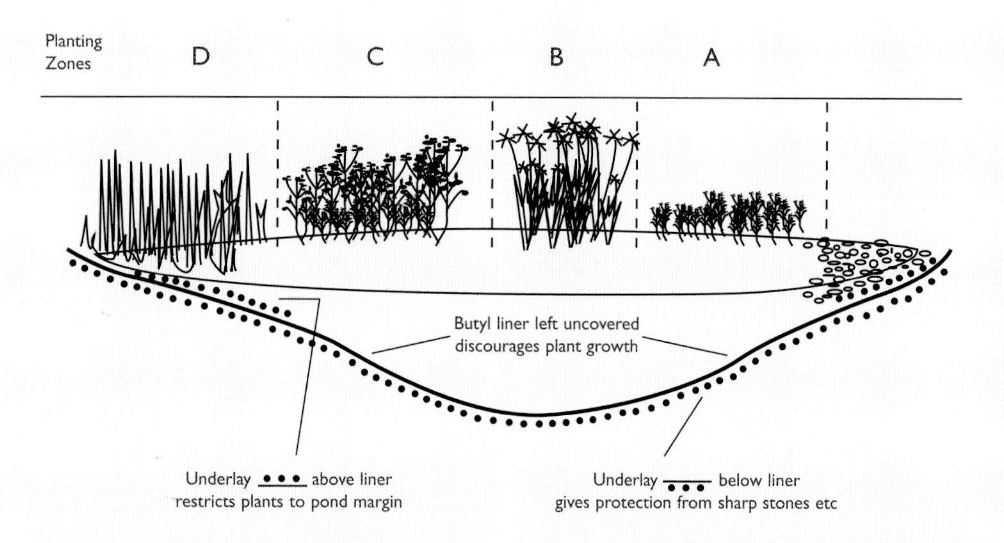

Marsh marigold is especially vulnerable to competition. Ragged robin and common valerian need space to seed since they are short lived. Brooklime spreads rapidly across mud when it appears but may struggle if the water level remains constant. Betony and devilsbit scabious need damp grassland near the edge of the pond, which can be cut at the end of the year.

Zone B – Deep-water specialists – 25 cm (10 in), uncompetitive
Although specialists of deeper water these plants can still be crowded out by plants from Zone C so this is where some annual management is needed. Flowering rush is a particularly spectacular plant.

Water plantain	Arrowhead
Flowering rush	Greater spearwort

Zone C – Shallow water 5-10 cm (2-4 in), competitive
This group of plants needs to be kept away from Zone A. They are tough, competitive plants so management is minimal.

Fleabane	Bogbean
Great burnet	Meadowsweet
Sneezewort	Hemp agrimony
Angelica	Greater birdsfoot trefoil
Gipsywort	Purple loosestrife
Water mint	Yellow flag iris

These plants are survivors and need little or no management: yellow flag iris (left), bogbean (right)

Deep water specialists such as greater spearwort need 20 cm (8 in) of water depth to reduce competition

Water plantain

Common valerian is an uncompetitive aquatic plant. It has a powerful scent in the evening

opposite: Less competitive plants such as marsh marigold need protection from too much competition

Bogbean anchors itself in the margin providing a floating raft over the water, which forms a vital protective cover for tadpoles, newts and other amphibians.

Zone D – Marsh 15cm (6 in), competitive

This group spreads by underground rhizomes. Zone C can be used to keep them in check. Their dense growth is ideal for nesting moorhen and general cover.

Branched bur-reed	Pond sedge
Common rush	Soft rush
Hard rush	Meadow rue

Given the above groups, the simplest profile for the pond is a saucer for all zones with two exceptions: Zone D requires a shallow area of about 1m (3 ft) in width, and Zone B requires deeper water of 20-30 cm (8-12 in), which would mean that the liner needs to be 50 cm (20 in) deep.

After the liner has been laid, the geotextile underlay should cover the planting areas in order to protect the liner from any sharp stones in the back-filled soil, as well as to limit the amount of water surface to be covered by vegetation.

PLANTING RATES

The rule of thumb for planting densities is to aim for approximately 4 plants per metre (3.3 ft) of pond perimeter.

If 9cm (3.5 in) pots are used, this gives a well-stocked pond margin needing very little management. However, if plugs are used, much more attention is required because the dormant weed seeds emerge from the seed bank and compete with the small plugs.

MANAGEMENT

The management of a pond is not demanding. The marsh plants in Zone D will gradually invade Zone C so either pull them up or use a herbicide.

Unlike the other habitats, ponds seem to have attracted more than their share of plants that must be avoided.

Canadian pondweed floats on the water surface and will cover your pond in a couple of seasons. If it is around on ponds in your area, moorhen or mallard will probably bring it. You will need plenty of time to rake it out.

Our native bulrush is another menace that will come in on the feet of waterfowl. It has roots like hawsers and must be removed before it gets established. In addition, do not on any account plant yellow loosestrife. This is another attractive plant, which will simply take over the pond margin.

SUMMARY

A pond can be excavated, planted up and fully functional within a single season.

- Butyl liners and clay are the two usual methods used to retain water.
- Pond design includes planning where to put the spoil from the excavations, fitting the pond to the landscape, and designing the planting zones.
- plants can be divided into four zones:
  - Uncompetitive, water's edge
  - Uncompetitive, deeper water
  - Competitive, shallow water
  - Competitive, marsh.
- Management is minimal, but marsh plants will attempt to spread out to other areas and mallard and moorhen will bring in unwanted plants such as bulrush.

Three competitive plants, which should have a place in every pond:
left: Hemp agrimony; **centre**: Purple loosestrife; **right**: Meadowsweet

A dense stand of bulrush, which has spread by under-ground rhizomes.
Although it provides excellent cover, it will fill up a small pond in no time

Muntjac escaped from Whipsnade Zoo in 1921 and are having a serious effect on our woodland flora

CHAPTER 13

RESTORING WILD FLOWERS TO WOODLAND

THE RE-ESTABLISHMENT OF OUR WOODLAND FLORA MAY WELL GO DOWN IN THE HISTORY OF THE COUNTRYSIDE AS ONE OF THE GREAT MISSED OPPORTUNITIES.

Significant areas of land have been planted with trees but if the grant-aided woods on our small farm are anything to go by, we shall all end up with rows of bare poles (young trees) with the wind whistling through them and no wild flowers or wildlife in sight. The rides are too narrow and quickly turn into dark corridors and the shrubs usually fail because they are either shaded out or destroyed by deer. A visit in May reveals all. If there is no bird song, questions need to be asked.

It is true that the shrub component in grant-aided woods increased to 20 per cent in the 1990s but this had little more than a cosmetic effect around the woodland margins. Foresters want to grow trees with straight stems and that means trees which are close to each other and which can be thinned in later years. For many a small farm wood of an acre or two, trees with shrubs (say one-third trees, two-thirds shrubs) are much better for wildlife but these trees grow bushier without such straight

Oaks dominate our valley landscape, but there is not a young tree in sight

pages 158-159: The almost completely dry month of April 2007 resulted in this wonderful display of bluebells and greater stitchwort continuing for weeks on end

We are now planting edge shrubs in this wood somewhat after the event because thanks to the Higher Level Stewardship Scheme we already have a deer fence

Choose a red fescue cultivar and an effective ground cover can be established cheaply. It excludes almost all weeds, needs virtually no management and provides cover for the small mammals that barn owls require

Many grant-aided woods are bare and draughty and useless for wildlife. Ours are no exception

stems, so what is a gain for wildlife is a loss to the forester. There is a conflict here. To make matters more complicated, deer re-appeared in the countryside. Apart from small areas where there were always fallow deer, roe and muntjac did not exist until the 1990s. Historically, there were sufficient people engaged in agriculture to ensure that if a deer appeared, it was in the pot in no time. Today, the countryside is deserted and the deer have it all to themselves This has raised the stakes massively against shrubs and coppice, which are the two obvious methods of improving small woods made up of nothing more than bare poles. Whereas you could guarantee 1.2-1.5 m (4-5 ft) of growth in a year from coppiced hazel or ash, today you are likely to end up with 12 cm (5 in). In some woods, rabbits are doing as much damage as deer.

It is significant that the richest habitat for our woodland flora has been coppiced woodland, particularly hazel. Re-establishing this with the appropriate woodland flora will be challenging at the very least and deer and rabbit protection will have to be taken very seriously.

But there is something even more important if we are to re-establish our ground flora, particularly on ex-arable land. We have to learn that although planting trees and shrubs on fertile arable land is fine, doing nothing with the land between the new trees is simply not an option, because it will make it ten times harder to introduce woodland flora. It is quite a thought! Fortunately there is a simple solution.

PLANNING THE GROUND COVER

Even before the design stage, planning needs to take into account the question of ground cover. If the new woodland is going to be on grassland with clover and rye grass, there is no problem. When the trees grow up, they will shade out the existing grass and clover and leave the ground bare for woodland plants.

The problem arises on ex-arable land, which we discovered to our cost and which is where the majority of new plantings are located. It is also rather strange that there does not appear to be a single organisation interested in the re-establishment of our woodland ground flora. Like many others, we planted our new small woods on ex-arable land. There was all the excitement of marking it out with a 3 x 3m (10 ft x 10 ft) grid and getting all the trees in with the new tree shelters and wondering if they would come through the dry spring. The trees flourished and despite a certain amount of unhelpful pruning by our neighbourhood roe deer all seemed well.

Then we began to notice the luxuriant growth of creeping thistle, docks, nettles, ragwort and much more. There was hardly a weed that was not represented, so many an hour was spent with the knapsack sprayer getting the place back under control. Our first plantings were in 1982 with two more small woods in 1990 and 1991, both of which required a great deal of work with the knapsack sprayer. When we extended the woods in 1995, the penny had well and truly dropped that we needed effective ground cover to prevent the weed invasion. By this time we had met numerous other farmers and landowners who had had similar experiences.

The solution was incredibly simple. After the harvest was over and when the stubbles were being cultivated, the intended planting area was included and then we sowed one of the red fescue cultivars, chewings fescue, at 25 kg per hectare (22 lbs per acre). Of course it got a certain amount of trampling during the planting of the trees but the ground cover was in place.

It is salutary to summarise the actions we then took or did not take:

Results of establishing red fescue ground cover

Years 1 and 2. Knapsack sprayer with herbicide round all trees for two seasons. Note that it is vital for vegetation to be a reasonable height so that it collapses on the ground and provides cover against other weed seeds blowing in. Knapsacking too early in the season can provide a fertile seed bed around each tree, particularly for spear thistle.

opposite: Woodland flora is adapted to long periods in the shade with a burst of growth when the light is let in. Parts of the woodland canopy should always be open

164-165: Bluebells have a remarkable ability to set seed in heavy shade

A wide ride on the Broomhill Estate in Dorset, where a charity, Will Woodlands, established a wild flower strip with impressive results for insect diversity

Year 2. Spot spray spear thistles in plantation. This was not a serious infestation, but had to be dealt with.

Year 3 and beyond. No further knapsack work around trees, and no cutting or mowing of grass.

What we did not realise with our earlier plantations was that further damage had been done because the weed seeds from the invading weeds were all now in the ground waiting to surprise us (again!) at a future date.

This duly happened. When we thinned out Oak Tree Copse a few years ago and let the light in, a massive number of nettles sprang up. Since the aim of the operation was to devise methods of re-establishing ancient woodland flora, we found ourselves back at square one with the knapsack sprayer trying to get on top of the weeds.

WOODLAND DESIGN

I am assuming that the choice of tree species has been worked out. In our case, we were very keen to get more oak back into the landscape, since these are the trees that dominate our valley and no one had planted any for over 150 years.

However, the planting opportunity goes much further than just planting trees. If the aim of the woodland planting is to increase bio-diversity such as birds, butterflies and other insects, you could include the following:

Edge shrubs to provide shelter on the edge of the wood and along rides.

Wild flowers in glades and rides to provide nectar.

Woodland flora to be established within the wood as soon as there is enough shade.

Edge shrubs. Before discussing the establishment of wild flowers in glades and rides, the issue of shelter inevitably arises. The wild flowers will not attract insects if the site is so windy that they get blown away! We can speak from bitter experience, since our valley is exceedingly draughty and the wind blows straight through some of our new woods. The establishment of shrubs, particularly on the side of the wood where the prevailing wind blows and along some of the wider rides is a very high priority. But how do you achieve this if there are deer?

It is no good putting shrubs in 60 cm (24 in) shrub shelters. The deer will eat them as they grow out of the shelters at a convenient browsing height. The most effective method is to buy some lightweight deer fencing and put it around a circular or linear group of shrubs for 5-6 years so as to get them away. The fence can then be moved along to protect another group. If there are deer around, it is wishful thinking to expect shrubs to grow without some sort of protection.

Wild flowers in glades and rides. The key point if you are considering establishing woodland wild flowers in new woodland is to use **meadow wild flowers for the first fifteen years**.

It is essential to wait until there is enough shade if you are trying to establish woodland wild flowers. If you jump the gun, the grass will dominate and all your efforts will have been wasted.

During this period, make the most of meadow wild flowers in the new rides using meadow wild flower seed mixes (see Chapter 7 for sowing methods). If the site has had a recent arable crop, you must establish a stable ground cover, so there is every reason to sow a fescue grass seed mix with a mix of wild flowers. If the site already has an established grass/clover crop, I would urge you to have a go at establishing wild flowers. Management through topping or making hay can proceed along the rides as if it was a meadow.

Woodland flora. After fifteen years the canopy thickens, the light reduces and the grass diminishes. The way is clear now to establish woodland flowers within the planted area and on the edge of the rides. One of the best environments is an area of hazel, which can be coppiced on a 7-12 year rotation, or longer if required. This ensures that the worst of the weeds are shaded out as the coppice grows and you get off to a fresh start whenever the area is coppiced.

Here again, protection from deer will be required.

You may be surprised to discover that there are as yet no agreed methods for establishing woodland flora. So what is the best method? There are severe constraints as to what can or cannot be done in a woodland area working in between hazel clumps or trees:

The ground cannot be prepared mechanically except with a hand-held rotovator, so all the preparation that has been discussed for meadow areas will be very much more difficult. If seed is contemplated, it will be impossible to mow, although hand strimming would, in theory, be possible. In any case, there are significant woodland plants that set very little seed.

If 9 cm (3.5 in) pots are used, the cost of planting up reasonably large areas en masse will be prohibitive.

This is where we need to fall back on the ability of plants to get on with life rather well on their own, so long as the shade levels are adequate and the weeds are kept at bay. Our trials are at an early stage, but we have planted out a range of woodland plants in 9 cm (3.5 in) pots in groups of five about 30-50 cm (12 to 20 in) apart. The groups themselves can be as much as 5 m (16 ft) apart. We got rid of any clumps of nettle and severe patches of creeping buttercup, and left the plants to it.

The results of the trials have been revealing.

Seeders

Dog violet. Our initial 5 plants reduced to 2; there was probably too much shade.

Foxglove. The plants have spread well from seed, with 20 plants after two years. Because foxgloves are biennials, we need to establish the alternate year.

Herb robert. This has seeded all over the place and is abundant in an area of 10m x 10m (30 ft x 30 ft).

Primrose. Our 5 initial plants look good, but there is no sign of any seedlings. Primrose is a slow coloniser.

Sanicle. Only 2 plants survived and there is no sign of any colonisation, but sanicle is another slow coloniser.

Stinking iris. Birds remove the large seeds, so it is anyone's guess where they will carry it and it will appear.

Wood avens. This is an exceedingly vigorous plant that seeds like mad. Just two plants in the group would have been more than adequate.

Wood spurge. Several plants have survived, but it is a tricky plant to get established and one that is short lived.

above: This tree and shrub planting scheme at Hampshire Butterfly Conservation's Reserve at Magdalen Hill Down included wild flowers sown after the trees and shrubs had been planted on the ex-arable land

opposite: The woodland trial area after the canopy had been thinned and the hazel "layered" to increase its density

Wood sedge. This has seeded really well and 3 plants in the group would have been adequate. (We kept red campion out of this tria because it seeds so prolifically.)

Runners

Bugle and sweet woodruff. 3 plants in both these groups would have been more than adequate, but given their strategy of spreading by runners only, more groups will be needed.

Yellow archangel. Only two plants survived, so reinforcements will be needed.

Given the characteristics of woodland areas, I believe that the careful use of groups of 9cm (3.5 in) pots is the best approach.

Bluebells

We have carried out seeding trials in several areas with a rate of 2 grams per sq.m (0.075 oz per 10 sq.ft), but this appears to be an absolute minimum. It is a great advantage that bluebell seed will grow in the deepest shade, so there are no concerns about weed competition, but please note that you will have to wait at least five years before you see any flowers. Trials with bluebell bulbs suggest that 30-50 bulbs per sq.m (10 sq.ft) is a reasonable rate at which to plant.

The one point to bear in mind is just how dynamic the woodland environment is. In a wet year, trees and shrubs may grow twice as much as in a dry year, so shade levels will increase dramatically unless some of the growth is cut back. Most of our woodland plants will move around and grow where the light is. Problems occur when the wood gets dark and light is even unable to get into the rides. Some plants such as germander speedwell will suddenly produce incredible displays of colour when conditions are ideal for them. Others, such as nettle-leaved bellflower, change very little from year to year.

Note on deer

Fallow. Re-introduced by the Normans after becoming extinct during the last Ice Age.

Results of the woodland trial

Primroses are slow to colonise

Foxgloves seed well if there is a pool of light

Bugle spreads steadily by runners; here with herb robert

Wood sedge seeds well if it is given enough light

Herb robert spreads quickly in semi-shade

The trial area beginning to look quite natural

Nettle-leaved bellflower in a wood on the chalk colonises with difficulty owing to its reliance on seed

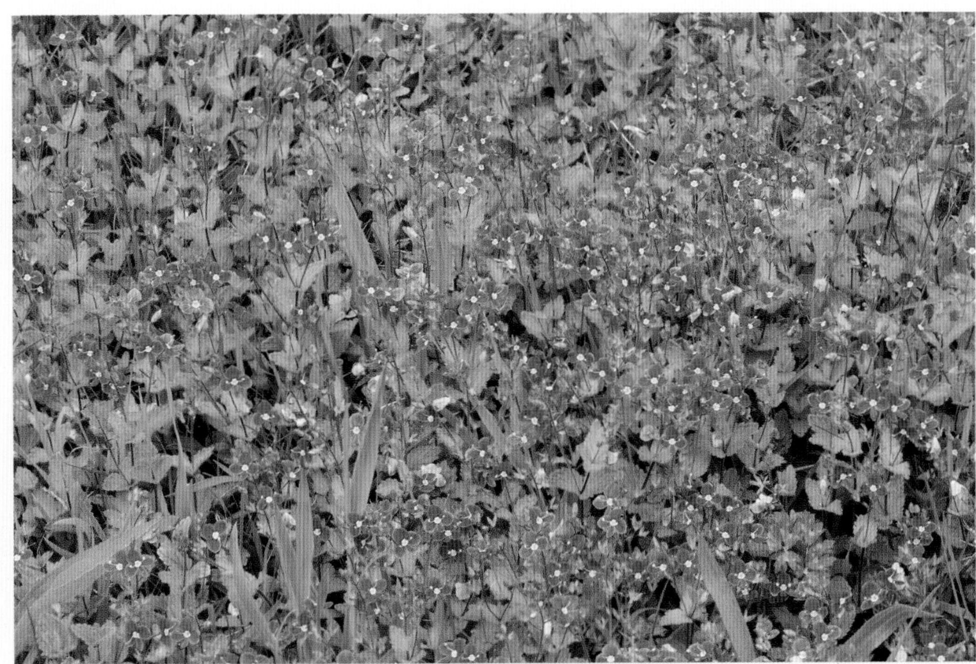

Germander speedwell – a superb display in a lane, the result of perfect conditions for expansion from runners

Roe. Migrated back to Britain after the last Ice Age. Protected by the Normans but in 1338 were declared "beasts of the warren" (unworthy of noble hunting) and became a food source for the expanding peasant population. By 1600 they were scarce but were introduced into parks and wildlife collections in the 18th and 19th centuries and have increased ever since.

Muntjac. The Duke of Bedford introduced muntjac to Woburn Park in the 19th century but the most significant release was from Whipsnade Zoo in 1921.

SUMMARY

Re-establishing our woodland flowers may go down as one of the great missed opportunities.

- There is a conflict between growing trees for foresters and woods for wildlife.
- Planting trees on ex-arable land requires precautions to be taken against arable weeds.
- Woodland design involves edge shrubs, meadow wild flowers and woodland wild flowers.
- Trials to establish woodland wild flowers suggest that groups of 9 cm (3.5 in) pots may be an appropriate method.

An unexpected visitor in our woodland trial area suggests that diversity is on the increase

CHAPTER 14

THE RESPONSE OF BUTTERFLIES TO A LITTLE HELP

WHEN YOU ARE WORKING AWAY TRYING TO ESTABLISH WILD FLOWERS FOR THE FIRST TIME, IT IS QUITE DIFFICULT ENOUGH WITHOUT LOOKING ROUND THE CORNER TO SEE WHAT MIGHT HAPPEN NEXT.

pages 172-173: Provide the right kind of nectar and along come the customers such as this Red Admiral

When we started sowing wild flowers, few people had tried it and most reported unfavourable experiences. Perhaps my expectations were low, so when the wild flowers in Cowfield actually grew, I could not believe that there would be excitement on another front. This was probably my mindset as I was walking up the hill one May evening in 1993 beside our new woodland, Middle Copse, where the trees were a few feet high with lots of tussocky grass in between. The sun was dipping below the ridge and only the top of the plantation was still catching its rays. As I approached, I noticed fluttering in a clump of tall cocksfoot grass. I looked closer and there were a dozen Common Blues going to roost. As the shadows crept up the cocksfoot stalks, the butterflies fluttered up as high as they could to get the last remaining warmth of the sun; and then, having got as far as they could, they were still. Seconds later the sun had gone. There was another group of Common Blues going to roost a little further on. I was mesmerised. I had never seen more than a

opposite: We can honestly say that Common Blues, seen here going to roost, are now common again because their food plants – birdsfoot trefoil and black medick – are now common on the farm

above: Given a hedge with some long grass and wild flowers you will have Hedge Browns

above: Small Skipper – rough grass and plenty of nectar are the simple demands of the skipper family

opposite: Small Coppers lay their eggs on common sorrel; if your farm or garden does not have any, they may be absent

Fleabane, an excellent late summer nectar source

Wild marjoram is not very palatable so the flowers survive well into the autumn

Devilsbit scabious will still have flowers in October

couple at one time, and I had no idea that this sort of thing happened.

There is something very special about wild flowers and butterflies in abundance. When Cowfield, our best meadow, is in full flower with Meadow Browns, Hedge Browns, Marbled Whites , Common Blues and Skippers all fluttering about, our visitors are entranced. When we tell them that they are standing in what was until quite recently the middle of a very large cornfield, they stare in disbelief.

But it shows what can be done with any piece of arable land or grassland if you set out to create favourable conditions, with the appropriate shelter, nectar and food plants. There is also the big "if" as to whether the butterflies you want will be able to colonise. Although we have much to learn, if a species is not very mobile and not in the locality, it is unlikely to arrive on its own. We have been remarkably fortunate at Shalbourne. What is so exciting is that there are great opportunities for anyone with a bit of land and imagination.

Because our whole farm was a couple of very large cornfields until some twenty years ago, I can explain what we had to do in order to bring back the common butterflies as well as many of the scarcer ones. It was all relatively simple: no need for degrees in biology. An additional factor in all this is that there are only sixty species of butterfly on the British list and you would be very unlikely to see more than forty of these on a lowland chalk farm such as ours. So the numbers are manageable.

ESSENTIAL CONDITIONS TO ATTRACT BUTTERFLIES

Provide high quality nectar

There seem to be remarkably few wild flowers which are really important for nectar although a great many more will be quite important. This is very much a feature of the natural world and is a huge help to us because it makes the whole task so much easier. A good supply of nectar keeps butterflies on the wing, in the same way that petrol keeps cars on the road.

Key nectar species

Birdsfoot trefoil	Yarrow
Ox-eye daisy	*Small scabious*
Field scabious	Fleabane
Lesser knapweed	*Wild basil*
Tufted vetch	*Wild thyme*
Devilsbit scabious	Cowslip
Greater knapweed	*Wild marjoram*
Salad burnet	Thistles sp.

NB Italics = plants of chalk soils

opposite: Meadow Browns are arguably our commonest meadow butterfly; our meadows have hundreds of them

In a typical seed mix there are other wild flowers that have useful nectar but the above are the most important.

Lesser knapweed is particularly helpful because, once established, it will continue to flourish even if the management is not very good. Note also that six of these species are plants of the chalk, which at once makes the point that anyone on chalky soils has an unfair advantage, which should be maximised. I have also included tufted vetch because it is so beloved by the Skipper butterflies. All these wild flowers grow well from seed, so there are no problems on the propagation front. It is always worth looking out for plants that grow well locally, but if most of your local wild flowers have been destroyed, you will have to search further afield.

Nectar all through the season. There is a powerful reason why you should consider more habitats than just the wild flower meadow. If you created a meadow and then cut it all for hay towards the end of July, where would all the butterflies go? Butterflies are on the wing until the end of August and often well into September. Of course you could leave a 1 m (3 ft) strip round the edge of the meadow, but it reinforces the point that you need wild flowers in many more places: along the hedges and field margins, beneath trees, along the lanes, and round the pond or in wet places. This brings us to some key wild flowers that flower late and have the capacity to keep a great many insects on the wing right through until September. I shall mention three of them.

The first is *fleabane*, a plant that spreads well by underground rhizomes in damp pastures, although it is quite happy in drier conditions. Its name derives from its common use in medieval times to help rid houses of fleas and their allies. Armfuls were collected early in the morning while the dew was still on the flowers and strewed throughout the house. As the temperature rose, the fleas emerged from the bedding, cushions, etc. to feed on the strong nectar. The staff quickly gathered up the fleabane and ran for the door and the bonfire!

The second key plant is *devilsbit scabious*. All the scabious family are excellent nectar species. Devilsbit scabious is again a plant of damp pastures but it is equally at home on chalk downland. It has become scarce because so much grassland has been ploughed and drained. It flowers well into October, which significantly prolongs the supply of nectar. It can also spread rather well by seed.

Finally, there is *hemp agrimony*, which grows well round ponds as well as on the chalk and, again, it is very attractive to butterflies.

If you provide nectar, how many species of butterfly are you likely to attract? There is a vital point with respect to food plants. A butterfly food plant (or larval food plant) is where the butterfly lays its eggs. Some 18 species of butterfly lay their eggs on widely available plants such as grasses or stinging nettles, so if you provide plenty of nectar, these butterflies should increase significantly. The species are listed below:

Butterflies with common food plants

Eggs laid on native grasses, such as cocksfoot, annual meadow grass, fescues	Eggs laid on stinging nettles	Eggs laid on cocksfoot Yorkshire fog, timothy	Eggs laid on brassicas	Eggs laid on crucifers
Meadow Brown	Red Admiral	Large Skipper	Large White	Orange Tip
Hedge Brown	Small Tortoiseshell	Small Skipper	Small White	
Ringlet	Peacock	Essex Skipper		Green Veined White
Marbled White	Comma	Speckled Wood		
(Wall)				
(Small Heath)				

opposite: Male Adonis Blue. If there is plenty of nectar about, butterflies will stay on the wing longer

As far as butterflies are concerned, perennial rye grass is useless.

Two other butterflies should be mentioned here.

A *Wall* has been seen on the farm but only very occasionally. This is a classic example of a butterfly with a population that fluctuates dramatically. For some reason the population is at a low ebb at present, so there is every reason for its scarcity.

The *Small Heath* is a very common butterfly of fine-leaved grasses in old grassland so why is it not present on the farm? It is unable to travel across the arable desert that starts at the foot of the Downs. It is a weak flier, hardly moves around at all, and has been unable to reach us. As soon as it does colonise, I would expect its numbers to build up quickly.

above: Green carpet moth is found in hedgerows and chalk grassland; it flies at dusk in June

opposite: Brimstones, seen here feeding on thistle, may fly up to a kilometre (more than half a mile) to find a suitable bush of buckthorn on which to lay their eggs

Provide key food plants

We have so far discussed eighteen species of butterfly, which have undemanding food plants and are fairly easy to please. Now it becomes more difficult. We have approximately the same number of butterflies (seventeen) that lay their eggs on specific wild flowers. I have divided them into three groups in the table below.

It goes without saying that if a butterfly has a food plant that is difficult to grow and to manage, we shall have a problem.

Butterflies that use one or two food plants:		
Common Blue	Brown Argus	Dingy Skipper
Green Hairstreak	Grizzled Skipper	

Butterflies that use tree/shrub food plants		
Brimstone	Holly Blue	Purple Hairstreak

Butterflies with a single food plant		
Small Copper	Small Blue	Chalkhill Blue
Dark Green Fritillary	Painted Lady	Clouded Yellow
White-letter Hairstreak	Brown Hairstreak	
Duke of Burgundy Fritillary		

Butterflies with one or two food plants: Intensive farming with herbicides on arable land will have removed all broad-leaved weeds (wild flowers) so if we want particular food plants we have to re-instate them.

The Common Blue. This used to be very uncommon on the farm. It is now widespread because its food plants – birdsfoot trefoil and black medick – are back in strength. Birdsfoot trefoil is a standard ingredient of all our wild flower seed mixes, so it is present in all our meadows and field margins. Black medick is common in the arable seed bank and so is now common on tracks, particularly on the chalk. Both species are easily grown from seed.

Brown Argus. We recorded this butterfly for the first time in 1998, but it may have been present earlier, the food plant being dovesfoot

above: Chalk scrape with well established horseshoe vetch

right: Kidney vetch is easy to grow on chalk banks, seen here at Magdalen Hill Down near Winchester; it is odd that no one has done more to encourage it

opposite: You can only describe the courtship of the Small Skipper as hectic

cranesbill, which grows as a common weed amongst our wild flower seed crops. The seed crops themselves provide the nectar. Common rockrose is another favoured food plant, which we have yet to establish on the farm.

Dingy Skipper. This was first recorded in 2001 in a sheltered glade in one of our new areas of woodland above the village. Once again birdsfoot trefoil is the food plant although greater birdsfoot trefoil is also used. This butterfly is not yet widely distributed on the farm but its arrival was very exciting.

Two other butterflies will, I hope, turn up at the farm in the foreseeable future. The *Green Hair-streak* likes bushy places and lays its eggs on dogwood, gorse, broom and common rockrose. The farm is getting gradually bushier, so I am optimistic. The other butterfly is the *Grizzled Skipper*, which likes a range of wild flowers such as wild strawberry, tormentil and agrimony. We are going to provide these plants, all of which are at present very scarce. I live in hope that both these butterflies will be able to colonise.

Butterflies with trees or shrubs as food plants: There are three species of butterfly on the farm that use trees or shrubs rather than wild flowers. Since trees and shrubs are quite easy to establish, it is well worth considering the needs of these butterflies when planning a countryside restoration scheme.

Brimstone. This butterfly is the harbinger of spring, beating up and down hedges on the first sunny days. It lays its eggs on buckthorn. Purging buckthorn is a useful constituent of any new hedge, and should be included. Although

opposite: Chalkhill Blue is one of the most beautiful butterflies of chalk grasslands but it needs horseshoe vetch in order to breed

right: The chalk pit where Chalkhill Blues will probably colonise from 2.5km (1.5 miles) away (circled)

no good as a hedging plant, alder buckthorn can be planted as an alternative plant for egg laying on acid soils.

Holly Blue. This is one of my favourite butterflies since it is easy to identify because of the way it flies high among bushes in gardens. But it has a complicated lifestyle using the flowers of female holly (or dogwood) to lay its eggs on for its first brood and then switching to the flowers of ivy for the second. Ivy will flower when it has climbed to the top of its support, be it tree, post or wall. The needs of this butterfly remind us again that holly should be present in new hedges as well as in new woodland plantings. The same goes for ivy, which often has a bad press. The truth is that diversity is everything in the natural world and however much you dislike ivy, it is important for this butterfly and for all kinds of other insects that use it as one of the scarce evergreen plants which are so valuable for hibernation. So by all means cut it off your favourite trees but make sure enough remains around.

Purple Hairstreak. This is a real crick-in-the-neck butterfly, since it flies about in the tops of oak trees and lays its eggs high up on the ends of the branches. We found we had a colony in a large old hedgerow oak, which was a big surprise. Oak will, I am sure, remain a key species for any woodland planting so this butterfly should be secure.

Butterflies with a single food plant:
Laying all your eggs on one plant would suggest a somewhat foolhardy strategy, yet this is what a number of our most beautiful butterflies do. It works well when the plant is common and easy to grow. When the opposite is the case, the status of the butterfly can get precarious. This is why organisations like Butterfly Conservation are of such value. The research they undertake is important in pointing out deficiencies in habitat where specific conservation action is required.

Small Copper. It seemed logical to begin with the one butterfly that is in no danger. The Small Copper is a really active little butterfly, which uses only common sorrel or sheep's sorrel, the latter being more of a heathland plant. When a new meadow is created in the autumn, it is quite usual for there to be Small Coppers prospecting the sorrel plants the following spring. It is remarkably quick off the mark!

Small Blue. This is our smallest butterfly and a really enchanting little insect. Its flight is somewhat weak and fluttering but this belies its ability to seek out pastures new. It arrived at the farm in about 1991 because we had been sowing kidney vetch, its food plant, in field margins and in Cowfield. Because kidney vetch is so palatable, it will not persist in grazed pastures, but it continues to grow very well on field margins where there are thin chalky soils. This is a butterfly somewhat under threat, so it is remarkable that so little has been done about it. Kidney vetch grows well from seed, it colonises well and any chalk roadside cutting or bare chalk surface would be an excellent site to establish a colony.

Chalkhill Blue. Providing for this butterfly is as difficult as the Small Blue is easy. Its food plant is horseshoe vetch, which will not grow from seed under field conditions, and does not readily grow from cuttings. The only way I can get it to grow is to cosset it in seed trays, when a good plant in a 9 cm (3.5 in) pot can be

achieved in about a year. I now have evidence that seedlings develop from planted out plants under field conditions, but it may only happen in certain years when the weather is just right, and the number of successful seedlings is tiny. Planting it out as a small plug does not work either because it grows too slowly to get through the first winter. To make things worse, the *Adonis Blue* also completely relies on this plant. So if ever there was a butterfly food plant where more research is required, this is the one. I have a special interest in the Chalkhill Blue because it has yet to arrive at the farm. It breeds happily in a small cutting in the chalk down above the village of Ham, 2.5 km (1.5 miles) away as a butterfly might fly. Tantalisingly, I have been told of Chalkhill Blues being seen in the village. We have established a small chalk area with 30 plants of horseshoe vetch that are growing well, and other areas of 25-50 sq.m (270-540 sq.ft) are going to be prepared. I am confident that we shall win this one.

Dark Green Fritillary. This is a magnificent butterfly which turned up at the farm for the first time in 1997, and again in 2004 when two adults were seen. Its food plant is the hairy violet, which needs quite close grazing of the vegetation to avoid being outgrown. I cannot speak from any experience of this plant since we have not yet tried to introduce it into one of our chalkier areas of pasture, but I think it might grow well. It divides up easily and comes from seed.

Duke of Burgundy Fritillary. I think it is unlikely that this butterfly will ever reappear at the farm, but we should, in theory, be able to provide for its needs. It appears to be declining, not because its food plants – cowslips and primroses – have declined, but because they are not being managed correctly. Management of these two Primulas has to ensure that they produce large, ungrazed leaves, which is what the caterpillar requires. However, with cowslips so easy to establish, there must be scope for extending the range of this wonderful butterfly.

The White-Letter Hairstreak. This is a fascinating butterfly which had as its food plant both the hedgerow elm, which made up much of our landscape, as well as wych elm, which was

above: *Painted Lady*. It is always exciting when you attract migratory butterflies, especially if they have come all the way from Spain or North Africa

opposite: *Comma*. Named after the letter C on the underwing. A butterfly of woodland glades and rides

right: *Small Tortoiseshell*. This is a butterfly that has declined massively in recent years but the causes are as yet unknown

more of a woodland tree. The farm landscape was dominated by the elm until Dutch elm disease struck in the 1960s. We still have hedgerow suckers trying to grow year after year. Could there be Hairstreaks still around? We need to search. It is a tantalising prospect.

The Brown Hairstreak. I always thought that this butterfly was far too rare for our farm until I met Henry Edmonds, mentioned earlier, who farms the Cholderton Estate not far to the south of us. He has wonderful blackthorn hedges, miles of them, with Brown Hairstreaks to match. I am developing blackthorn hedges on our farm. Surely, one day we shall have them too!

There are two additional butterflies that occur at the farm and that should be mentioned – the *Painted Lady* and the *Clouded Yellow*, both of which migrate from the Continent and arrive in May and June since they cannot usually survive our winters. However, we see Painted Ladies in far greater numbers, so the Clouded Yellows always give rise to more excitement. Of course nectar is needed for both but they are undemanding with their food plants: Painted Ladies require thistles and mallows; and Clouded Yellows require clovers, lucerne and birdsfoot trefoil.

Join up butterfly hotspots with wildlife corridors

We have made a number of meadows and glades, which I will refer to as butterfly hot spots because they have had such a dramatic effect on the numbers of butterflies. Butterflies are attracted by sunny and sheltered spots that are the antithesis of huge open cornfields. Our smallest hot spot – one of the best places to watch butterflies on the farm – is at the bottom of a new, small woodland, where the last 20 m (60 ft) was not planted up but sown to a grass/wild flower seed mix. It is particularly sheltered – butterflies hate being blown about. It is tiny, 20 m x 25 m or 500 sq.m (just over a tenth of an acre or two-thirds of a tennis court).

Before you rush off to convert your tennis court to wild flowers let me emphasise that although these areas are of value on their own, they are of hugely greater value if they are joined together by field margins, wildlife corridors, nectar strips, call them what you will. An increasing number of farmers and landowners are using 6 m (20 ft) strips of wild flowers to join up these hotspots.

above: This glade of wild flowers, which is smaller than a tennis court, is one of the best places on the farm to see butterflies

Connecting species-rich habitats on a Hampshire Farm

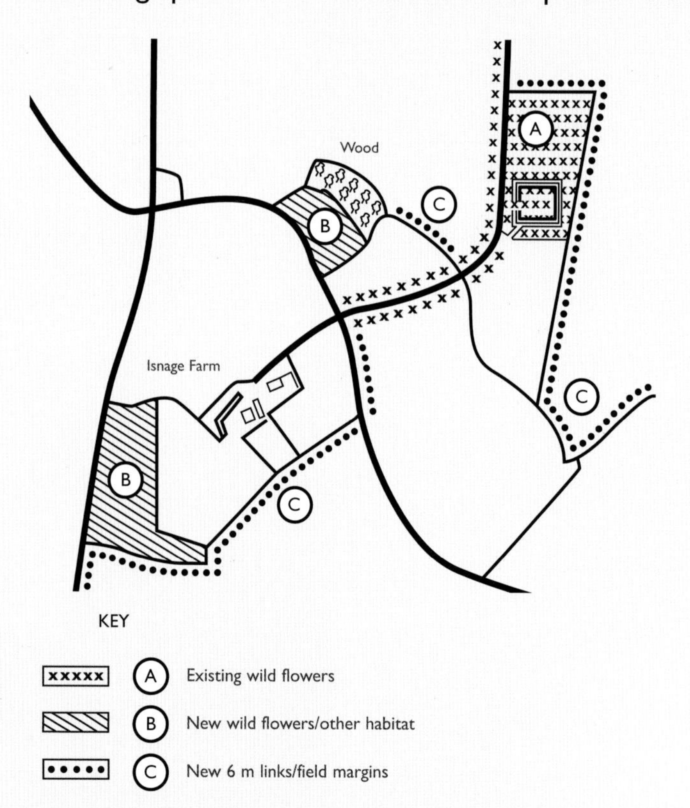

Small isolated fragments of species-rich habitats will lose diversity unless they can be linked together. Rowan Downing's farm has well established as well as newly established wild flowers and these are now linked together by 6 m field margins

opposite: Wild flower field margin on Rowan Downing's farm in Hampshire

Colonisation of Marbled White butterfly at Winterbourne, Berkshire

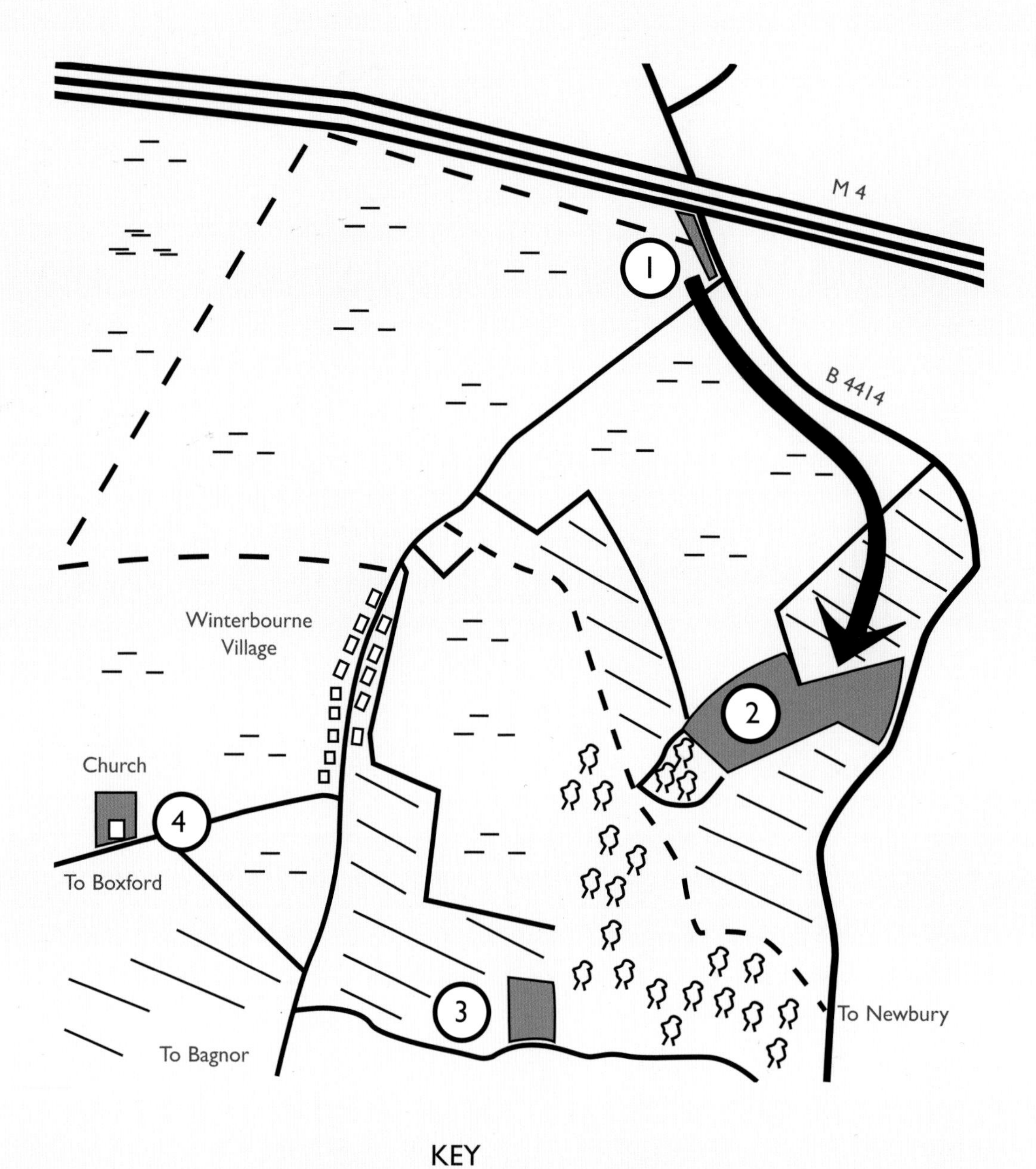

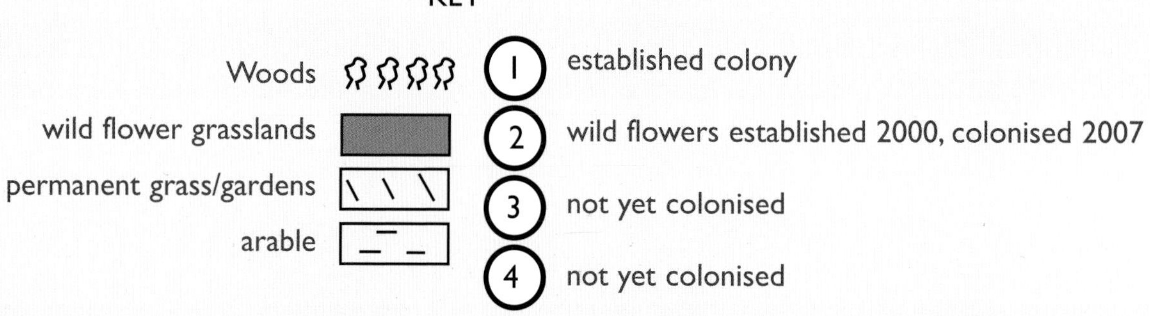

Arrow indicates route of colonisation along roadside verge and through gardens

The Marbled White

This is a classic case history relating to the need for field margins. In the village of Winterbourne, just to the north of Newbury, the Marbled White is beginning to re-colonise lost ground, but only with great difficulty. It is quite a large butterfly and appears to fly strongly, so it should colonise easily. This is not so. It has always had a colony along a roadside verge and on a section of the embankment beside the M4. Four years ago a 4-hectare (10-acre) field was converted to wild flowers a few hundred yards away. The butterfly was there in strength last year. However, exactly one mile away there are two perfect sites where it has not been seen in living memory. One is to the south-west in the churchyard, where there are plenty of wild flowers, and the other to the south where a new meadow was created some twenty years ago. The butterflies cannot get there because there is, in each case, a large arable field barring the way. If there were some field margins, it would have had no problem.

Given a network of 6 m (20 ft) nectar strips, the whole exercise of providing for butterflies and insects becomes much more cost effective. You do not need wild flowers everywhere but can concentrate your efforts on growing and managing them in the areas where the conditions are most favourable to them and then link up these areas.

Record what happens, however briefly

It is so important to record, however briefly, what you do on a farm or in a garden to encourage wildlife. You never know how things are going to turn out and when something that you have initiated works out exceptionally well, it is essential to tell others so that they can benefit from your experience. If you never write anything down, it makes it harder to help.

It is also valuable to record how new butterfly colonies develop. For example, although *Small Coppers* seem to be able to colonise effectively, they do less well when the sward thickens up and competes more with sorrel, their food plant. Similarly, *Brown Argus* do well on dovesfoot cranesbill but since this is an annual, it dies out when the sward becomes established, so where this is going to happen, you need common rockrose, a perennial. With the creation of an increasing number of species-rich grasslands on ex-arable land, lessons are quickly being learned on the selection of key species and their management. Learning how to encourage the establishment of ant colonies, which have important associations with some of our blue butterflies, is proving difficult and information on this topic is desperately needed.

Our farm had a tiny area of old grassland banks of less than an acre beside Carvers Hill, the road that leads down to the farm. The banks were too steep for the fertiliser spreader and they supported an old chalk grassland community of plants with a number of butterflies, which Jack Coates, a local butterfly enthusiast had recorded over the years before I purchased the farm. These records were most valuable, particularly as our big experiment in Cowfield was just across the road.

So that is as far as we have got with our butterfly experiments. In summary we have: 18 species with undemanding food plants, of which two are occasional visitors; 12 species with more demanding food plants, of which two are occasional visitors, which makes a total of 30 species seen on the farm. I live in hope that we can persuade a further 5 species to colonise.

SUMMARY

Wild flowers provide high quality nectar. If they can be provided together with the plants on which the butterfly lays its eggs (food plants), the return of butterflies will exceed all expectations.

A short list of wild flowers, which are easy to grow and can provide high quality nectar through the season is included.

A number of individual butterflies are discussed:

- 18 species which lay their eggs on grasses or stinging nettles.
- 5 species which have several food plants.
- 3 species which have tree or shrub food plants.
- 7 species which have a single food plant.

Very small areas of wild flowers (500 sq.m or about a tenth of an acre) can be incredibly valuable so long as they are joined up by 6 m (20 ft) field margins or wild flower corridors.

pages 194-195: Brown Argus probably lays its eggs on dovesfoot cranesbill or other cranesbills as there is no rock-rose at the farm

CHAPTER 15
THE RESPONSE OF BIRDS

A VISIT TO A FARM OR GARDEN IN MAY OR JUNE WHEN BIRDSONG IS IN FULL SWING IS A GREAT TIME TO ASSESS HABITAT QUALITY. BIRDS ARE AT THE TOP OF THE FOOD CHAIN, SO A HEDGEROW FULL OF SINGING YELLOWHAMMERS INDICATES SUCCESSFUL NESTING, WHICH MEANS THAT THE HEDGE IS IN GOOD SHAPE AND THAT THERE ARE PLENTY OF SEEDS OF WILD FLOWERS AND OTHER PLANTS.

If there is a deafening silence in a particular habitat, all may not be well. So how are we doing at Carvers Hill Farm and what have we learnt? Here is a brief review of the birds and the habitats we have been discussing: meadow, hedge, pond and woodland. In most cases we only have indirect evidence that the existence of wild flowers has been beneficial, but if plant diversity has increased, it will certainly have benefited birdlife.

MIXED CROPPING (ARABLE AND MEADOW)

Skylark. When Pond Field was converted from corn to grass in 1993 a pair of skylarks nested successfully two years later for the first time for many years. We now have three pairs, so something is going right. Skylarks feed their chicks on insects and spiders for the first week of life. This surely must be helped by wild flowers in the field margins, but it may equally have been the organic corn crops. Our main wild flower meadow, Cowfield, was probably too dense to be used by the young. With an imminent change back to conventional spring cropping, we shall be able to assess the importance of the organic crops. Will the introduction of conventional spring crops and winter stubbles outweigh the move away from organic crops? It is a complex situation.

Lapwing. These returned in 1997 to breed in our wild flower seed crops where there was an abundance of insects, but after a couple of years they disappeared. It could have been the neighbourhood fox or the fact that the area simply was not open enough. They returned again and bred successfully in 2007, which may

Spring crops and wet meadows are ideal for lapwings

opposite: Kestrels are at the top of the food chain. Their presence is an encouraging sign

pages 196-197: Flock of golden plover coming down at dusk in winter to roost on an arable field. They breed in upland Britain

This glade is buzzing with insects. It must help the food supply for insect-eating birds

One of our new hedges, which has been colonised by yellowhammers

Male yellowhammer. Good hedges, preferably with a field margin full of insects, are vital for the successful rearing of chicks

With reasonable cover in small broad-leaved woods, there should be plenty of blackcap and chiffchaff. We have some way to go

have been due to our mini wetland being full of water. The move back to spring corn should provide ideal conditions and with more grazed pastures with wild flowers available, there will be areas for the chicks to feed.

Grey partridge. How do we restore its fortunes? Many farmers and landowners have provided ideal hedges, field margins and adequate predator control to little effect. Other factors such as pheasants laying their eggs in partridge nests may also be important but any additional nectar is bound to help insect life. The really major change since the partridges declined has been the change from autumn to spring sowing and the loss of all the winter stubbles with their huge source of winter food. So with these three species, it is the farming system which may be more important than the provision of wild flowers.

HEDGES

Some 3 km (2 miles) of hedges are now being restored, many of them with a wild flower field margin alongside. Although the hedges with their new hedge trees are transforming the farm landscape, it is the wild flowers at their base that have the greatest impact on wildlife, but without detailed research it is impossible to quantify.

Yellowhammer. The "little bit of bread and no cheese" call of the yellow hammer is as typical of farmland as the song of the skylark but more discreet. While we have been restoring the hedges, their population has increased from two or three pairs to about six. It will continue to increase as the hedges improve, but our 6m (20ft) field margins of tussocky grass or wild flowers are also contributing to nesting success. They provide seeds for the adult birds as well as insects for the chicks.

We can also influence yellow hammer numbers by helping them through the winter. Wild flowers no doubt provide seeds in the autumn but it is the provision of winter food that transforms the winter food supply, and some of the birds attracted by the food may stay on to breed. The term "wild bird seed" is a new phrase that has come into farming parlance. The Government's new Environmental Stewardship Scheme launched in 2005 has a range of options for the encouragement of wildlife-friendly farming including the provision of 0.5 ha (1 acre) plots of wild bird seed mixes for wintering birds. This is proving most effective. I was visiting a farm in the early winter of 2006 when I saw a 1 hectare (2-acre) field planted to a mixture of kale, quinoa and other seed-bearing plants with a large flock of over two hundred chaffinches, yellowhammers and reed buntings (yes, reed buntings!). The neighbouring farmer was so impressed with this great flock of finches that he reviewed a somewhat unproductive 2-hectare (5-acre) field of his a couple of hundred metres away, and decided to sow it to the same mixture. What a gain for wildlife!

Linnet. The UK population has generally been on the increase but it was still cause for celebration when linnets returned to breed in 1997. Since then we have seen a build up of winter flocks to well over a hundred birds with a probable three pairs now nesting. Linnets need thick thorny hedges, bramble and scrub to nest but since they eat seeds throughout the year our wild flower field margins will have made a significant contribution. However, the winter flocks may also be due to the wild bird food that we have provided.

Two additional hedgerow birds should be mentioned here: firstly the *whitethroat* where the numbers have fluctuated from 2 to 4 pairs. Wild flowers attract insects and this is of great value to these insectivorous birds, but it is the problems in their wintering area in the Sahel in West Africa that probably determine how many pairs we see on the farm. Then there is

Our wetland attracts coot (above) and mallard (right). You never know what may fly in

the *long-tailed tit*, which returned to breed in 1998 and has built up to 3 pairs using the larger unmanaged hedges. Here again, the increased insects from the field margins have been of benefit.

WETLANDS

We have attempted to reinstate water in three ways: permanent ponds, temporary ponds and grassland that floods in winter. The permanent pond beside the farm buildings was a significant investment because of the need for a butyl liner. It was planted up with the full complement of aquatic species to provide abundant nectar through the season. Once again, the effect of this nectar supply cannot be quantified but the insect life on the pond is visually good and the presence of 5 or 6 species of dragonfly bears this out. The *moorhens*, which dominate the pond, usually bring up two good families. So the pond appears to be in rude health.

The temporary pond in Pond Field is fed by a land drain, so all we did was to dig it out. As with the other pond, *moorhen* nest and *mallard* and *heron* visit, but everything is less successful here, apart from the frogs, which breed in their hundreds. This is what attracts the heron, which picks off quite large frogs. We have never tried to plant up this pond, since it seemed rather an opportunity to see how colonisation proceeded on its own, so there is a less good supply of nectar. But the deciding factor is the way the pond dries up in late summer, and many of the marginal wild flowers cannot survive these conditions apart from the growing clump of bulrush, which we shall have to do something about or there will be no open water.

The mini wetland was created by building an 80 m (90 yd) bund in Valley Meadow, which has impounded nearly half a hectare (an acre) of water in winter; it fluctuates with the rainfall and returns to grazed pasture in summer. This has brought in *green sandpiper*, *teal*, nesting *mallard* and *moorhen*, *coot*, *heron* and *wagtail*, as well as *swallows* and *house martins* feeding in summer. Quite apart from providing a marvellous additional habitat, it is a stunning landscape feature. At present wild flowers play no part in this. We intend to increase the nectar supply with late summer wild flowers such as fleabane, meadowsweet, betony and devilsbit scabious. It will be interesting to see what effect this has.

Another bird we have been seeing increasingly in recent years, but only in winter, is the *reed bunting*. The provision of more nectar and rough grass with plenty of seeds of all kinds should encourage them to stay on the farm and breed.

WOODS

As I suggested in an earlier chapter our small woods are not perhaps our greatest achievement in terms of wildlife, but there are bright spots on which we can build.

The older section of Oak Tree Wood (which was established in 1982) was planted beside an old hedge with mature trees in it, so we have *tawny owls* in this part of the wood, with *great spotted woodpeckers* feeding but not yet nesting. Nest boxes have had an excellent effect on *great tits* and *blue tits* by providing an alternative to the cavities that take time to develop in older trees. This is good as far as it goes.

But what about the shrubs and the wild flowers? They are mainly absent; if we want them, we shall have to plant them. The bramble that dominates in many a wood has yet to arrive although there are signs that it is beginning to creep in from the adjoining hedges. As a result there are few *blackcaps* and *chiffchaffs* although a pair of each seem to have survived in Oak Tree Wood, but I would describe their conditions as spartan. If we can create shrubby corners we shall attract *willow warbler* (which nests in a thick hedge at present) and *garden warbler*, which was recorded for a few years when the new plantations were growing up.

Our permanent pond close to the farm buildings has a dense margin, which has helped the moorhens breed successfully

top left: moorhen; **top right**: moorhen nest

Blue tits respond quickly to nest boxes

As described in an earlier chapter, it is the presence of deer that makes establishing shrubs so difficult. Middle Copse is now deer fenced, so the newly planted shrubs will be able to grow without constant pruning; and woodland flowers are being established in Oak Tree Copse. We know how well wild flowers grow in a woodland glade because Jack's Hill Copse on our northern boundary has a glade through the centre of it, sown to wild flowers some years ago, which is buzzing with insects. But once again establishing the shrubs has been impossible because of the deer, so there is very little birdlife.

What is exciting is that creating the right sort of habitat is bringing back the farmland birds. It was not surprising that the removal of half our hedges led to a 50 per cent reduction in the population of yellow hammer, so it is entirely logical that restoring hedges will reverse the decline of this delightful bird. Similar measures will reverse the decline in other species such as *skylark* and *lapwing*. There will inevitably be declines in some species, such as the migratory *turtle dove*, where there are factors out of our control.

MIGRANTS

When we try to improve habitat so as to increase the breeding success of farmland birds, it is easy to forget that winter survival is usually the limiting factor as to how many birds we start off with in the spring. Winter habitat can be as important as summer habitat although the reasons are different. For winter visitors who visit our farms, this is, of course, the only habitat that matters. These visitors come from upland Britain or from countries such as Scandinavia, where food is unobtainable for most of the winter. Their arrival gives an extra perspective to a farm or garden, especially if you are fortunate enough to be on a migration route such as a valley or ridge. Since many of them migrate at night on their epic journeys to Africa, you will not see them unless they stop for a day or two in order to feed. If you offer the right sort of food, they will stay longer. We now have records of 18 species that visit the farm in winter or pass through as migrants.

Our farm lies in the upper part of the River Shal, which runs from Shalbourne north to Hungerford to join the river Kennet, a distance

opposite: Green woodpeckers now breed on the farm because of the increase in permanent grassland

right: Hedges laden with berries provide vital food for flocks of fieldfares and redwings from Scandinavia

of 6.5 km (4 miles). Below Shalbourne, the valley is aligned in a south-west direction, with a parallel downland ridge. There is plenty of evidence that both these features are used as a migration route.

There are two groups of birds that concern us: winter visitors, which use the farm in winter for varying lengths of time depending on the weather and the food supply; and long distance migrants from Scandinavia or Africa, which pass through the farm in autumn or spring and stay for anything from just a few hours to a number of days.

Winter visitors. Most of our winter visitors are easy to observe, because they stay for weeks or months, depending on their food supply. The occasional *woodcock* comes from northern Britain or Russia and keeps to the woods, where shrubs or bramble are important to keep the ground free of frost. Woodcock need soft ground in order to feed with their long beaks. *Fieldfares* and *redwings* descend from Scandinavia and clean up the hawthorn berries in the hedges before moving on, so it is important that some hedges are left unmanaged for maximum berry production. *Green sandpipers,* also from Scandinavia, may be in the area for most of the winter and make the occasional visit to our wetland in Valley Meadow. *Snipe* from northern Britain call in when there is a spell of severe weather. Wild flowers do not play much of a role in any of the above, but *meadow pipits* from upland Britain and *skylarks* move through the farm in autumn and are attracted to the grass and wild flower seeds in Cowfield.

Long distance migrants. These birds are mainly insectivorous, which brings us back to wild flowers and their role in attracting insects. *Sand martins* on their way to West Africa are often seen in late August hawking for insects over the field where we have our wild flower seed crops. We see *redstarts* occasionally in May on their way back to upland Britain from south of the Sahara, and then in autumn we see traffic in the opposite direction with *wheatears* leaving upland Britain on their way back to Africa. *Whinchats* used to call in at harvest time on their way to tropical Africa, but we have not seen them for many years. However, we had a visit from a *stonechat* in April a few years ago, probably a local bird since they breed on the downs quite near the farm.

These migrants require insect food and shelter, so we need to provide hedges or green lanes along the valley with plenty of wild flowers with nectar, especially those with late nectar. This brings us back to the much-maligned ivy, which provides arguably the best quality nectar for insects in September and October.

Gilbert White wrote in his *Natural History of Selborne* in 1773 that wheatears were sold in vast quantities in Brighton and Tonbridge and "appear at the tables of the gentry that entertain with any degree of elegance." He also observed that he never saw them in flocks but always in twos and threes. We now know that they were on their way to Africa. Much needs to be discovered on a local level as to which valleys are used most and how we can help, but there is no doubt that food and shelter are what is required.

SUMMARY

With birds at the top of the food chain, their presence or absence from farm habitats is significant.

- *Skylark, lapwing* and *grey partridge* have probably been affected most by the move from spring sown to autumn sown corn with the loss of winter stubbles.
- *Mallard* and *moorhen* do well on the farm's two small ponds but it is the newly created wetland that is bringing in new species such as *coot, teal* and *green sandpiper*, and helping the *lapwing*.
- *Yellowhammer* and *linnet* have increased steadily as the farm's hedges have improved but a supply of winter food has also been important.
- Woodland species such as *blackcap* and *chiffchaff* are surviving in small numbers but we need to introduce a substantial shrub element in our woods if these are to increase.
- Habitat quality and diversity are just as important for migrants, with wild flowers in sheltered corners crucial to provide insect food for species such as *redstart* and *wheatear.*

opposite: The robin population has increased in line with more hedges, thickets and small woods

CHAPTER 16
GIVING WILD FLOWERS A HELPING HAND

I AM IN NO POSITION TO JUDGE WHETHER I HAVE BEEN ABLE TO CONVEY TO YOU THE SERIOUSNESS OF THE POSITION IN WHICH WILD FLOWERS FIND THEMSELVES, BUT IF WE DO NOT IMPROVE THE SITUATION, THE REMORSELESS DECLINE OF OUR WILDLIFE WILL CONTINUE.

Plants that were once common are becoming rare and formerly rare plants are becoming endangered. However, what has now been demonstrated by many organisations and individuals is that if the right action is taken, recovery of the particular wild flower and the insects that depend on it can be swift. You might ask what the Wildlife Trusts were doing (each County has one), but the simple answer is that the Wildlife Trusts were so under resourced that they were and still often are struggling to keep up with managing their own nature reserves, saving threatened habitats, advising planners, and a myriad of other demands that are made upon them.

Since 1989 there has been a new conservation charity called Plantlife, dedicated to saving plants and, although quite small, it has proved effective at taking action and co-ordinating the efforts of others. One of its first tasks was to prevent extinctions. Protecting threatened species in their natural habitats required focussed action, which Plantlife has been able to bring to bear, especially in understanding more fully the species' ecological requirements. All this required resources and here again Plantlife has been able to tap into and encourage a great wealth of volunteer support and expertise. Over 250 sites are now managed for what are termed "back from the brink" species: these are either extremely rare species or threatened populations involving about 100 individual species; 55% of these are now stable or on the increase, an excellent example of what a carefully targeted approach can achieve.

The larger problem is preventing the common plants from becoming rare. Species with which many of us are familiar are becoming so scarce that we no longer see them. Each county is different in that plants that are a feature of one county may not be nearly so common over the border. The losses have been massive. In East Wiltshire two plants seem to epitomise this: meadow cranesbill is still common but has declined massively in recent years; Dyer's greenweed was always local, but has now all but disappeared.

The use of herbicides on farms, especially the use of Round-up, has virtually eliminated wild flowers from huge areas of the countryside, leaving our roadside verges as the only remnant of much of a county's flora. But even here, herbicides were used for several years to "control" the vegetation. Today's management is not much more encouraging.

Meadow cranesbill was common along miles of roadside verges. It isn't any more. The flail mower comes along at random times and more often than not cuts at completely the wrong time of year. The plant continues to decline. With payments to farmers for managing the countryside becoming more widespread, we now have the absurd situation of our farm being paid to restore wild-flower-rich field margins (with meadow cranesbill) on one side of a roadside hedge whilst the wild flowers are being massacred by the roadside flail on the other side. Surely we can do better.

Dyer's greenweed was local and rather rare. Most of the rough ground and grassy places that suited it have been ploughed and it was finally lost from one meadow where it survived until recently when the grazing regime became too intensive. I know of only a single site within 16 km (10 miles) of the farm, with the exception of Salisbury Plain, which is a marvellous exception to all the rules and where there is

pages 208-209: Eyebright exemplifies all that is difficult about some of our wild flowers if we want them to be common again. The seed is not commercially available so you have to hand collect it, and it only grows on thin alkaline soils where grass competition is minimal or where there is ample grazing. If you want wild flowers, do not start with this one!

opposite above: Meadow cranesbill was once a common flower but it has declined massively. It is seen here in a roadside verge with agrimony

masses of it. Yet this is a really attractive broom-like shrub which looks stunning when in flower and would be an asset to any garden.

With so many plants experiencing massive declines, any local genetic variations are being lost. Attempts have been made to protect roadside verges over recent years, but with frequent changes of contractors, it is no surprise that implementation of these arrangements has been patchy at best. Yet most people are unaware that there might even be a problem, which is remarkable considering how many of us live in the south of England. Is this a case of a massive failure of education? The problem lay partly in the fact that herbicides destroyed our flora so rapidly at a time when farmland or what went on there was perceived as very much someone else's problem. But of course it's a problem shared by all of us. It's our countryside, not someone else's. So what can be done?

RAISING AWARENESS

There is a huge awareness task to be tackled. Wild flowers are now so scarce in the wild that it may only be through restoration schemes that people are introduced to wild flowers at all. The scope for planting wild flowers in parks and other public areas is enormous. Without a major increase in awareness, no one is going to notice if the decline continues. If this awareness is going to start in schools, local farms in rural areas may have to restore small communities of some of the commoner wild flowers, otherwise no one will remember that they even existed. Somehow, and we are an inventive nation, we shall find ways in which we can involve young people in wild flower restoration schemes.

There is now a new awareness among farmers and landowners that wild flowers are important, not just because of their wildlife interest, but because they provide nectar for countless insects in the food chain and are also important in encouraging insect predators in their role as controllers of insect pests. But too many farmers are still blissfully unaware of the jewels of plant diversity which they have on their land. It would seem to be a high priority to encourage the custodians of our country-side, which is, of course, what the farmers are, to learn about and protect this incredibly valuable resource.

WHAT CAN THE LONE INDIVIDUAL DO?

There are no short cuts here. As I hope this book has suggested, the first task is to learn the commoner wild flowers, and then to keep your eyes open when you are travelling around your own patch of countryside, which is not always comfortable for your fellow passengers. But it is essential to know what is going on locally.

Then, if you feel that something needs to be done either with regard to protecting some wild flowers or to encouraging them, you can contact Plantlife, which can act as a useful source of advice, expertise or support. It also needs volunteers for its work and there are many ways in which you can contribute.

Plantlife also runs a "Grow wild to know wild" campaign, which encourages gardeners to learn more about the plants they are growing, to get out into the countryside to see them, and to join the fight to conserve them.

That seems to sum up very adequately what the interested individual should do.

The local school in action planting up the scrape at Hurst water meadows

The scrape beginning to fill

Results at the end of the season

GLOSSARY

Acid soils – where the soil water has a pH of less than 6.5.

Acres / hectares – 1 acre = 4840 square yards and 1 hectare = 10,000 square metres. The acre suits the small scale of many of our countryside features whereas the equivalent 0.40 hectare is fairly meaningless.

Alluvial – land subject to regular flooding during which silt is deposited.

Amphibians – cold-blooded vertebrates whose temperature varies according to their surroundings, and that have to resort to water to breed. Includes toads, frogs and newts.

Ancient woodland – woodland that has existed continuously from before about 1700.

Annual – a plant that completes its life cycle in one year, i.e. it grows from seed, flowers, fruits and dies.

BBOWT – Berks, Bucks and Oxfordshire Wildlife Trust, one of a network of Wildlife Trusts which cover all counties in England with the aim of conserving natural habitat, if necessary by acquiring nature reserves. See Wildlife Trusts.

Biennial – a plant that takes two years to complete its life cycle.

Broadcasting – sowing seed in such a way that it lies on the soil surface without being buried.

Butterfly Conservation – an organisation formed in 1968 with currently 33 nature reserves and 12,000 members. Committed to reversing the decline of butterflies and moths.

Butyl – flexible synthetic rubber material used to line ponds.

Calcareous – soil rich in calcium salts derived from chalk or limestone.

Chough – an uncommon but handsome relative of the crow with red legs and bill.

Coppice – underwood trees that are cut to near ground level and then grow again from the base. Also used to describe woodland which is managed for its underwood.

Copse – another word for "coppice".

Cornfield annuals – an endangered group of plants that were once plentiful in cornfields but which have been all but wiped out by modern herbicides. Unlike the annuals in a meadow, such as yellow rattle and common vetch, cornfield annuals require the land to be cultivated if they are to grow.

Countryside Stewardship / Environmental Stewardship – Environmental Stewardship is now replacing Countryside Stewardship as a means of encouraging farmers to deliver effective environmental management. Entry Level Stewardship (ELS) is open to all farmers, Higher Level Stewardship (HLS) being discretionary for farmers who are able to deliver significant environmental benefits in high priority situations.

Cut and laid hedge – an ancient technique of cutting a hedge after 10-15 years. The stems are only partially cut and are then laid in one direction, with vertical stakes knocked into the ground every half metre (20 in); the tops of the stakes are secured by flexible hazel shoots or bindings.

Dew pond – a shallow pond in a relatively waterless district, such as the chalk downs, which has a specially prepared bottom to catch and retain as much rain as possible.

Diversity – the number of different species in a particular area.

Domesday Survey – a detailed survey of all the land in England ordered by William the Conqueror in 1086.

Drill / seed drill – a piece of agricultural equipment which enables seeds to be planted in straight equidistant rows in the bottom of furrows. It was invented in 1730 by Jethro Tull, who farmed on the edge of Shalbourne. The land between the rows can then be hoed / weeded.

Enclosure period – Acts of Parliament that provided for the enclosure of land so that it could be protected from neighbours' livestock, as compared with the traditional open field system of strip farming. Main period of enclosures 1714-1800.

Fescue – several of our commonest, most valuable forage grasses belong to this genus, notably sheep's fescue (*F. ovina*) and Red fescue (*F. rubra*). These grasses can exploit some of our less fertile soils such as the thin chalk.

Flora Locale – an organisation that encourages the wise use of wild plants for native planting schemes with wildlife in mind. It provides advisory leaflets on all types of habitat restoration and runs a programme of countrywide workshops.

Forage crop – a crop grown specifically to be cut and conserved for later use.

Genetics – the scientific study of heredity and variation.

Habitat – the environment of an animal or plant, comprising the whole complex of vegetation, soil and climatic factors to which it is adapted.

Harrow – a metal frame with tines drawn by a tractor, the equivalent of raking on a large scale.

Hay meadow – flat, fertile areas of land which could grow good hay, the only source of conserved fodder that could keep farm animals through the winter months.

Hayrick – a circular stack of hay with a thatched top that used to be found in groups near the farm in what was called the rickyard.

Headland – the edge or margin of a field, where tractors turn and where crop yields are often reduced.

Heel in – the practice of digging a trench to temporarily cover young woodland or hedging plants with soil until they can be planted in a permanent location.

Hybrid – the progeny of animal or plant parents of different genera, species, subspecies or even varieties. Some of these progeny may display remarkable vigour.

JCB – J C Bamford launched the Company that bears his initials in 1945. The JCB is the most popular wheeled or tracked excavator in the UK.

Layering – a technique for increasing the density of hazel stools (clumps) by cutting two-thirds of the way through a stem and pegging it down on the ground to form a new stool.

Legumes – the seed pod of a member of the pea family (Leguminosae) but used by farmers generally for crops of this family, such as peas, beans and clovers.

Metric measurements – this can only be described as a complete mess. We take a pragmatic approach, using acres for small areas – see above, but kilograms are very useful if you are measuring seed and cannot handle the mathematical dexterity required for pounds and ounces. As a result we work in kg per acre with kg per hectare always available.

Nectar – a sugary substance secreted by flowers in glands known as nectaries to attract bees and other insects which helps the process of pollination.

Nitrogen – inorganic nitrogen is used by crops; any not taken up in the growing season is usually washed out by rainfall.

Nutting – the country custom of gathering nuts from hazel bushes for storage. The arrival of the grey squirrel has put an end to all that.

Open fields – the large unfenced fields around medieval villages farmed under the strip farming system, where individual farmers were allotted strips of varying soil types and fertility.

Parliamentary Enclosures – see Enclosure Period.

Pesticides – includes all chemicals applied by farmers, herbicides, fungicides, insecticides et al., a bewildering arsenal whose long-term effects on the environment are only now beginning to be understood.

Plantlife – formed in 1989 with currently 23 nature reserves. It aims to protect Britain's wild flowers and plants, fungi and lichens, and the habitats in which they are found.

Plug – a seedling grown in a cell of a certain size.

Pollard – a native hardwood tree cut at 8-12 ft (2.5-3 m) and allowed to grow again, a method of achieving a crop of poles and avoiding browsing (eating) by grazing animals.

Precision seed drill – a seed drill where rubber belts with holes that fit the seed size are used to place the seed at precise intervals in the row, hence precision drill.

Rhizome – a rootstock or underground stem producing roots and leafy shoots.

Runners – a creeping stem above ground that roots at the tip to form a new plant that may eventually become independent of its parent.

Sahel – a strip of semi-arid grassland stretching 2,400 km from the Atlantic Ocean to the Red Sea, which varies in width from several hundred to several thousand kilometres.

Scallops – an edge with curves.

Seed bank – a collection of seeds that survives in the soil for a number of years.

Set-a-side – a system whereby agricultural land is taken out of production.

Share farming – a system of farming where ownership can be separated from the actual farming, with the owner contributing the land and the inputs and the farmer the labour. Some sort of deal is then done on the share of the resulting crops. This avoids the inflexibilities of a tenancy.

Silage – this is a revolutionary approach to conserving grass for the winter, which enables a farmer to cut a grass crop and then bale it while it is still green. This has been a major advance for the wetter parts of Britain, where making hay was a nightmare.

Spinney – officially this is a wood that consists of or has consisted of thorns, but it is often used to describe a small group of trees and shrubs.

SSSI – Site of Special Scientific Interest. There are 4000 sites in England covering 7 per cent of the land area. Over half of these sites are of international importance.

Stakes and bindings – see cut and laid hedge.

Stubbles – the un-harvested stems of an arable crop which often contains important winter food (wheat and barley in particular) for a variety of farmland birds.

Sward – a mixture of grasses and other plants, wild or cultivated.

Topping / mowing – these are two very distinct operations: topping is like a flail and chops up the grass into small pieces, which are then left on the ground. Mowing is a precision job, cutting the grass crop neatly at the base of the stem, so that it can be collected later.

Wading birds / waders – these have legs of varying length to exploit various water depths, with beaks also of varying length to exploit different food levels in the mud / soft grass.

Water table – the level below which fissures and pores in the strata are saturated with water.

Waste places – the term "waste" is an ancient one describing unenclosed land used for common pasture. This has all but ceased to exist.

Weed flush – the characteristic green haze of weeds that appears a few weeks after a piece of ground is cultivated.

Wildlife Trusts – a network of 47 Trusts, mostly formed between 1956 and 1964 with a total membership of 726,000 dedicated to conserving the UK's wildlife habitats and species.

INDEX

The occasions when a spark is ignited by a conversation or a meeting which then kindles a life-long interest, are but few. From an early age I had been a keen bird-watcher. It was my good fortune to know Charles Floyd, who was the first Chairman of the Wiltshire Trust for Nature Conservation (now the Wiltshire Wildlife Trust). He was concerned about my lack of interest in the plant world and so one spring morning he took me for a walk along the sparkling brook, which fed the moats at Great Chalfield Manor where he lived. The brook meandered through a long meadow, the meanders being marked by occasional scrub or bramble, so that you never knew what was round the next corner. As we approached the first bend, Charles Floyd observed, "We should find some self-heal just starting to come into flower." Sure enough, there were its lovely blue flowers just emerging. Then it was the turn of meadow vetchling. "We should be seeing its tendrils round the next corner." And there they were. After many more finds, we ended up with the leaves of Bath asparagus, a fascinating plant which is thought to have been brought to Bath by the Romans. I was mesmerised. How could anyone have such a detailed knowledge of a meadow? I was soon to find out that Charles's knowledge of natural history was encyclopaedic. But the spark was lit and my interest grew, and I have Charles Floyd to thank for a most precious gift.

ACKNOWLEDGEMENTS

The information for this book started with the very limited efforts at our farm and then progressed through hundreds of individual schemes on farms, estates and in gardens, so I thank all those who have contributed, often unwittingly, to pushing forward the boundaries of wild flower restoration know-how.

I owe very special thanks to my wife Hatty, who was the indefatigable proof reader and wise commentator, as well as to all the members of my family who frequently rescued me from computer glitches. That the farm has been a beacon of good practice in wild flower restoration is entirely due to the efforts of Bob and Rosanne Anderson. Of equal importance are our farming contractors the Stone family from Great Bedwyn, just over the hill. Gordon and his sons George and Andrew have been an incredible support in carrying out with unfailing good humour and courtesy what they must have often considered to be crackpot schemes.

I particularly want to thank the professionals, who usually appeared fortuitously at just the right time: Dr Chris Smith, who monitored our first field scale restoration scheme; Phil Wilson who found our corn buttercups; Jack Coates who was monitoring the butterflies on our farm long before I appeared on the scene; Henry Edmonds, who is a real pioneer in chalk grassland recreation; Dr Charlie Gibson, who set up the monitoring scheme for Butterfly Conservation at Magdalen Hill Down; and Keith Tomey, warden of Snelsmore Common Country Park, who gave invaluable help in identifying individual species.

Numerous individuals and organisations have allowed us to photograph their hard won restoration work in order to illustrate this book, and have assisted in other ways: Mary Baylis, Peter Booth, Robin Buchanan-Dunlop, Johnnie Buxton, the Charity Will Woodlands, the Churchwardens of St James the Less at Winterbourne, Rowan Downing, Lynn Fomison, Hampshire Butterfly Conservation, the Hurst Water Meadow Trust, Charles McGregor, Anthony Mildmay-White, the Ministry of Defence, the National Trust, Paul Parsons, Ian Pasley-Tyler, Oliver and Lucinda Steel, John Wheeler, John Wilmer. Grateful thanks are due to Dr Peter Pritchard and to Pete Potts for the use of their photographs, and for the Museum of English Rural Life at the University of Reading in allowing me to use six photographs from their archive. Taking photographs of wild flowers when the weather is against you is not for the faint-hearted and I should like to pay tribute to the remarkable photographic skills of Mike Bailey and Steve Williams.

Others need to be mentioned because of their support at critical moments in this odyssey, in particular Bill Acworth, Gerald Boord, Will Harley from Kennet District Council, The Organic Research Station – Elm Farm, Heather Ray from the former Countryside Commission, Alan Treasure, the Wiltshire Wildlife Trust and Wiltshire FWAG (Farming and Wildlife Advisory Group).

I am especially grateful to my publisher, the late Andreas Papadakis for his enthusiasm and drive in creating a book that is not only useful but succeeds in conveying the beauty of wild flowers; and to his daughter Alexandra for the design and her invaluable contributions to all aspects of the book; to Leyton Brown for his work on the layouts and to Hayley Williams for her meticulous care with photographs and illustrations.

PHOTO CREDITS

All photos © Mike Bailey and Steve Williams except for the following:
Pages 11, 12, 20 (bottom), 21, 22 (bottom), 28 (top left, centre, right and bottom left), 34, 35 (bottom), 47 (top right), 50, 55, 56, 67, 70 (left), 79, 82, 88, 89 (right), 90 (left), 101 (left and right), 106, 115 (left), 119 (right), 120 (bottom right), 135, 139 (left, centre and right), 141 (right), 143 (bottom), 166, 168, 170 (top left), 178 (centre), 185 (right), 190 (top), 200 (top right), 201, 203 (bottom) and 207: © Charles Flower. Pages 16 (bottom), 20 (top), 22 (top), 23 (top) and 32: © The Museum of English Rural Life. Page 45: © Keith Tomey. Pages 80 (top, centre left and right) and 81 (centre): © David Fenwick, www.aphotoflora.com. Pages 96 (top left) and 211 (bottom left, centre and right): © Dr. Peter Pritchard. Pages 96 (bottom left and right) and 97: © Pete Potts. Page 196: © David Broadbent, rspb-images.com. Page 199: © Ray Kennedy, rspb-images.com. Page 200 (bottom): © Steve Knell, rspb-images.com. Page 203 (top left): © Sue Tranter, rspb-images.com.

We gratefully acknowledge the granting of permission to use these images. Every possible attempt has been made to identify all images and contact copyright holders. Any errors or omissions are inadvertent and will be corrected in subsequent editions.